Jon Close is an English Solicitor with over 12 years' experience specialising in construction & engineering law, bringing particular contractual expertise in the procurement of energy, infrastructure, precision engineering and cyber security installations. He works daily with standard form contracts (JCT, NEC, ICC, FIDIC, IChemE, IET MF, ACA) plus a myriad of bespoke forms including funder-led EPC and EPCM for energy usage both nationally and internationally. In addition to this, Jon is a seasoned construction litigator and an accredited commercial mediator, which he firmly believes helps inform how he deals practically with risk when drafting contracts.

A former head of department and partner at a regional law firm in the South West for 9 years, Jon joined new law firm pioneer, Keystone Law, in March 2016 where he continues to act for a number of well-known clients, predominantly in the UK construction market. Married with a young son, Jon lives and works from his native Gloucestershire whenever professional commitments permit him to do so.

A Comparative Guide to Standard Form Construction and Engineering Contracts

A Comparative Guide to Standard Form Construction and Engineering Contracts

Jon Close

Consultant Solicitor, Keystone Law

Accredited commercial mediator

LLB (Jnt Hons)

Construction & Projects

Law Brief Publishing

Published 2017 by
Law Brief Publishing
30 The Parks
Minehead
Somerset
TA24 8BT

www.lawbriefpublishing.com

Paperback: 978-1-911035-32-9

To my wife, Antonia, for her unfaltering support as both a fellow lawyer as well as my closest friend and to my young son, Sebastian, for his unceasing wonder at the world he inhabits and his brutal honesty as one of my most fiercest of critics.

Contents

INTRODUCTION

As in many areas of life, the more choice that you have, sometimes the harder the decision can be to make. The same can apply to choosing which construction or engineering contract is most appropriate for a particular project.

In the United Kingdom, there is a mind-blowing array of contracts to choose from published by bodies such as the Joint Contractors Tribunal, (JCT) the New Engineering Contract, (NEC) La Fédération d'Ingénieurs-Conseils (FIDIC) Institution of Mechanical and Electrical Engineers (IET) and the Institution of Civil Engineers (ICE) as well as the Association of Consultant Architects, (ACA) Association of Consulting Engineers, (ACE) Royal Institute of British Architects, (RIBA) and the Chartered Institute of Building (CIOB)– to name just a few.

The busy contracts manager or other professional would be forgiven for simply sticking to what they know best or which someone else in the office has used previously (and many do so with the best of intent). The problem is akin to one faced with trying to put square pegs in round holes. Sometimes you may be able to force one through which roughly reflects what is due to happen (or more often, what is currently happening) on site, everyone signs it to keep the insurers/lawyers/funder happy and then assigns it to a bottom drawer somewhere or equivalent. Only when someone does not get paid or there is a question over variations, design responsibility or the existence of defects does the contract ever see the light of day again, by which point memories have faded, the parties are focused on umpteen other live matters and all parties are scratching their heads muttering, "what on Earth is this contract all about again?!".

I am not going to lie to you – there are plenty of cost consultants and lawyers out there who make a decent living out of the above scenarios occurring (me included) however in over more than a decade of legal practice I have seen countless clients resign themselves to the fact that this is simply how it is done and the cost of recovery is just something they need to factor into annual budgets. Yet, it seems that this global acceptance that 'this is how things are' is not embracing best practice, creates more cost than is strictly required and wastes an awful lot of management time which could, to be frank, be better used elsewhere.

So, the motivation for writing this book stems largely from seeing the frustration of others in firstly not translating accurately to the drafters of contracts what they want to achieve and their priorities and, secondly, not understanding (whether by design or simply lack of time) what options exist out there to help deliver those goals and then understanding some of the basic mechanics of the clauses which exist in order to give certainty and enact in a structured fashion any change or variation from the initial brief once things get underway.

We can never completely eliminate risk but within the plethora of options available, much exists already out there to help us minimise things going wrong. When they do go wrong, it is rarely because the relevant clause of a standard form contract does not give an answer; it is normally down to a number of human behavioural factors which falls outside the scope of this publication.

This guidance note does not seek to replicate guidance published by the various institutions and bodies themselves or to critique certain clauses as they are currently drafted. I recognise that there are many variables in play in any given set of circumstances which may lead to a particular decision being made. Rather, this guidance seeks to act as a reference guide for all users of such contracts as an entry level (or aide-memoire) document for the busy construction professional and their advisers who may be working outside their specialist sector. Nor can I compare and contrast too wide a range of standard form contracts here. If I did, then the purpose of this quick reference would be lost. There is scope however for someone grouping contracts into their sector usages and doing so in that logical way. Perhaps, this entry guidance will motivate someone to do so in the future as there is most definitely a need.

What I have attempted to do is remind the reader of the series of thought processes they should look to go through to help identify which contract, from three of the most commonly used industry forms in both construction and engineering, is/are most likely to help achieve project goals when faced with certain specific criteria[1]. To that end, it may be useful to refer to the various bodies' guidance publications in tandem with this document.

Please note that contract references throughout this guidance refer to those contracts used in England and Wales; not Scotland or Ireland which have similar, but not identical provisions.

By understanding the appropriateness of each form and the available conditions in relation to one another, I hope that this note will serve to minimise project risk and further the achievement of the 'Holy Trinity' of completing on time, to specification and to budget. Seeking this achievement is a running theme and the focus throughout the guidance note.

Before deciding upon a particular standard form of contract, it is important first to establish the procurement strategy. Therefore, the first part of this guidance note explores what users really mean when they say they are 'managing risk' and the range of procurement options open to them to address those identified risks and other project factors.

This is aimed at addressing the common and unsatisfactory scenario whereby the procurement route is decided before the client brief has been fully established, resulting in certain post-tender reviews of documents once the specific risks are apparent. This can have a negative impact on the parties' relationship from an early stage where they become entrenched and end up engaged in long, drawn-out and costly contract negotiation.

The guidance seeks to make contract users aware, or to provide comfort by way of reminder, that risk is inevitable but that it is not something to fear provided a strategy is put in place for handling it in one way or another. Possible strategies could be to:

- transfer it contractually;
- mitigate it by way of insurance or other avoidance measures;

1 ACA PPC2000 and TPC 2005 (as amended) etc. are referred to when appropriate for illustrative purposes.

- accept that it is there and put processes in place to minimise the impact upon the organisation; or
- do nothing at all.

The aim is to take the reader to a stage between Parts 1 and 2 of the guidance whereby it has identified the risks associated with certain client objectives and then s/he needs to look at how particular contract forms can help him or her implement a project to meet the three key deliverables noted above.

In the second part of this guidance note, therefore, readers are reminded of the range of contract options which mirror particular procurement routes and then consider which operative clauses from each of the suites the reader would then focus upon to deal with time, quality and certainty of completion.

As is often stated, one size does not fit all. Therefore, ultimately, it is hoped that this document will assist the reader in the compilation of contracts with the fundamental points above having been at least considered, if not all three being achieved to a greater or lesser degree. The guidance which follows here considers how a reader may possibly approach the decision-making process with reference to the main forms of contract in use from each of the bodies and some of the more global issues that they may face.

As I mentioned above, the risk of signing up to contractual terms which the parties never intended to be in place, either through a lack of understanding or failure to appreciate their significance, can and often does lead to a mismanagement of expectations on all sides. This is the seed for discontent which can lead to expensive disputes in terms of time and cost which should be minimised as far as possible. Hopefully, this guidance note goes some way to assisting in that goal.

In summary, my intent has been to make this as simple as possible by distilling the process down into four key questions which I, as someone who drafts contracts like you, would need to consider further; namely:

- What are the client drivers for my project?
- What are the procurement routes available to me?
- Which contractual structures encapsulate my chosen procurement route?
- Which clauses are most important to me in drafting the contract?

I sincerely hope that this book assists you in your daily drafting activities. All mistakes within it are solely attributable to me, for which I apologise for any in advance. As ever, this reference guide is no substitute for seeking proper legal advice when you deem it necessary. Hopefully, whenever that occurs, you will have even more of an insight into the logic and mechanics of drafting as a result of reading this book.

Good luck in all your professional endeavours and I sincerely hope that this publication makes your life just that little bit easier as a quick reference guide in the ever increasingly, fast-paced contracting World we all inhabit.

Jon Close
Stroud, Gloucestershire, UK
April 2017

PART ONE

WHAT DOES THE CLIENT WANT TO ACHIEVE?

1.1
KEY DRIVERS OF THE POTENTIAL PROCUREMENT STRATEGY

The procurement strategy identifies the best way of achieving the completion of a construction project – often taking into account the best value for money over the entire life-cycle of the building or facility.

The aim of a good procurement strategy is to achieve the optimum balance of risk, control and funding for a project. The choice of a particular procurement strategy largely depends on a client's required balance of cost, quality and time risks.

It is worth remembering that the construction project itself may only be a relatively small part of the entire life-cycle of the building or facility as a whole. The procurement strategy for the construction project therefore also needs to take into account where and how the construction project sits in relation to that wider picture. The issue of 'sustainability', particularly in respect of public sector works, is a major factor that needs to be considered at this stage.

To establish the procurement strategy, the procurement team needs to establish what the client's attitudes and key drivers are, and needs to state them in a way and in a language that is understood by the entire project team. More often than not, the lawyer (or whoever is going to reduce this concept into a contractually binding agreement for tender etc.) is not party to discussions on the basis that they are purely commercial considerations.

In my experience, this is a flawed strategy as many a commercial nuance has been lost in translation or simply omitted when contracts are then drafted. Whilst it may appear to add upfront cost to a project, you actually close the gap and avoid contractual risk being placed with the wrong party (which is often a cause of extreme irritation) by considering commercial and legal considerations together. Trust me, you will end up with a better draft the first time around (saving cost and time) if everyone knows what the contract is supposed to look like and its feasibility, as opposed to reverting to the square pegs in round holes strategy which creates endless waste and further frustration.

Fundamental to the process is understanding what the client requires from the project, together with some key information as to how much it wishes to spend, what quality it is seeking to achieve, and by what time it requires the project to be completed.

This process may take some considerable time, but this is of vital importance given that that there is so much that flows hereafter. It is usual, in many cases, for the procurement team to assist the client with the formulation of their own 'brief' for the project as, initially, it may well be that the client does not know much about the construction project. This is of course perfectly understandable especially if the client is an end user of the facility, but it will know the output or outcome that it requires at the end of the process. This output or outcome requirement needs to be expressed in such a way that the construction team can understand the client's needs and desires.

Do not be shy in 'playing back' to a client what you believe they are trying to achieve. If you cannot understand what they are trying to do, then you risk producing a contract draft which is at best a template unrelated to the reality of what will occur on site and, at worst, is wholly inappropriate as a contractual tool, confuses everyone and gets dismissed as another expensive irrelevance. Keeping the outputs simple when the design, technology and other concepts are anything but simple is an art form and will help you distil what is really most important and provide clarity to your first draft.

For this purpose, it may be useful to provide the client with a selection of 'high level' construction project outputs which they can consider further, for example:

(i) a project at a stated cost;

(ii) a project of a certain size, shape, appearance and quality; and/or

(iii) a project to a defined timescale.

It is likely that a number of options and sub-options will also be provided (with suitably qualified statements to cover any assumptions made, so that the client is aware of the parameters upon which the guidance is being provided) but these are your foundations.

Eventually, the client's attitudes and drivers can be established through the above process, and it is these attitudes and drivers that are behind the choice of a particular procurement strategy and procurement route. These attitudes and drivers are usually considered against the headings of:

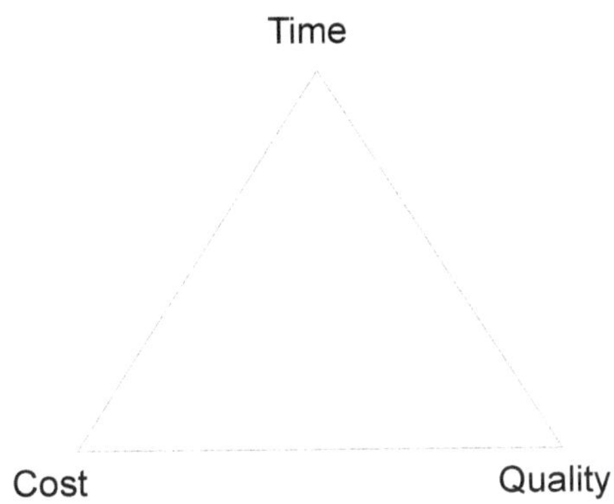

1.1.1 Cost

The cost of the project will be of paramount importance to the client, in that often a business investment decision is based upon the balance between the return or benefit to be achieved against the investment to be made. This is often considered as being two sides of an equation which must always balance each other out – with the surplus (as added value) being the difference between the resource input (the investment) and the benefits to be obtained. There may be a need to reduce construction project costs because of other cost pressures in the overall development spend profile.

Some of the questions that may be asked at this stage are:

- Is there a need or desire for cost certainty?
- Is a lump sum contract preferred?

- Will that be a fixed price or with fluctuations?
- Do the circumstances dictate remeasurement of the works?
- Must all tenders be on a competitive basis; are restrictions placed based on the form of funding available or required to be used?

1.1.2 Quality

Issues surrounding quality often go hand-in-hand with the extent to which a client needs (or perceives the need) to maintain control over certain aspects of the design and level of workmanship being employed.

Typical questions to ask are likely to be:

- Is the project for a completely new detached building, an extension to an existing building, the refurbishment of an existing building, the restoration of a historic structure or a repair and maintenance programme?
- Should the design be wholly in the hands of the client's consultants or can some detail design be placed as a contractor's responsibility?
- Does an innovative design demand sophisticated construction methods?
- Should there be provision for design by specialist subcontractors?
- What measure of control will the client wish to exert over materials and workmanship?
- How much reliance can be placed on performance specified requirements?

The answers to the above will assist you in terms of deciding whether other professional appointments are needed as well as the level of design sign off and change control that the client will expect to be in place governed, naturally, by the contract you will draft.

1.1.3 Time

Time is likely to be a very important factor for many clients and not just in terms of how quickly you can turn around the necessary contractual documents.

The financial benefits of undertaking the project in the first place will usually be based upon certain programme constraints. The projected financial outcome will therefore only be achieved in full if the project commences the 'delivery' of its benefits in line with a particular timescale.

Some of the questions that may be asked in respect of this matter are:

- Is a 'fast track' project with the shortest overall programme a priority?
- Is an early start date required; will there be adequate time to prepare full information for tendering purposes?
- Do circumstances dictate a specific completion date; can the contractor be provided with exclusive possession from its start on site?

1.1.4 Risk

As everyone is aware, risk is inherent in the construction process. The way that risk is to be dealt with is a fundamental issue when considering procurement strategies and procurement routes.

Any risk which serves to increase the costs and thereby reduce the 'added value' over the life of the construction project, or which has an effect upon the quality of the construction project that can be achieved, or which can delay the delivery of the construction project to the target timescale, will adversely impact upon the client's successful outcome of the construction project.

The risks may appear in any number of different ways, many of which are outside the direct control of the client. It may be forced in these circumstances to react to these risk events and to seek to change the scope, the standard, the cost or the time period for the project in some way.

Such risks can occur at any point in the life of the project and the earlier those possible risks are recognised and taken into consideration and allowed for within the chosen procurement strategy and procurement route, the less disruptive those risks will be if they actually occur.

The source of the risks might include some or all of the following:

1.1.4.1 Political or economic change (globally, nationally or locally)

A change in government and/or a change in the economic climate prevailing at the time may result in a different landscape upon which project decisions are to be made.

On occasion, this might result in the original project scope and size being delayed, rather than changed in scope, and then undertaken at a later date. An example of this may be a government decision to cut back on investment in a particular area in times of economic hardship, thereby putting the whole justification for the project in doubt. If contracting in regions which are prone to conflict or terrorist attack, the commercial risk is that a project could be suspended for a period of time. In turn, this could lead to a difference between prices originally tendered and what they subsequently are later on in real terms once the notice to proceed has been issued. If this is a real risk, then contracts with price fluctuation clauses will be at the forefront of your mind to allow flexibility in the event of transportation and goods/materials costs being so high that a party simply cannot proceed, meaning that the whole project falls over before it has begun. This can be avoided.

1.1.4.2 Legislative change - National or local government changes in policy or law

This is of particular relevance in industries where the client's affairs are highly regulated and where such changes would have a fundamental impact on the manner in which the client's business is conducted. An example of this may be Brexit in the UK where the introduction of new laws (or abolition of certain EU based laws) could restrict the ability of the client to conduct his business with the flexibility enjoyed previously and the need to adapt suitability to ensure continuing compliance. In such circumstances, contracts which do not rigidly define personnel on site as being the contract price paid but which recognises that people may need to be based away from the site but are equally still part of the cost may inform decisions as to the form to be adopted.

1.1.4.3 Environmental influences

Over the course of recent years, the trend has been to place a greater emphasis on environmental issues, thereby forcing firms to take account of such pressure in the conduct of their business and in the products or services offered. An example of this may be the need to amend production methods such that a greater proportion of the energy used is sourced from renewable resources and with sustainability of the environment in mind. Consequently, certain methods of collaborative working may need to be engendered within the contractual draft and/or various pre-conditions (or conditions precedent as they are often referred to) may need to be included in order to meet such expectations.

1.1.4.4 Social or technological change

A change in the social environment or some technological advancement may result in the client needing to amend his/her thinking behind the entire justification for the project.

An example of this may be the adaptation of a retail offering from an out-of-town development to one which offers further public transport infrastructure or is adapted to be moved to an inner town regeneration site with further physical restrictions and deemed liabilities in terms of site conditions. Or the proposed development of an incineration plant, the technology for which is significantly out of date but which due to funding restrictions needs to go ahead and with engineering installation contracts signed by a specific date. In such cases, a closely monitored re-design process could be accommodated subject to clearly worded post contract variation and valuation clauses being operated so that one party does not end up with a severely bad bargain (which then needs to be justified to shareholders or the tax payer) at the expense of another's windfall due to simply the passing of time and technological advances.

1.1.4.5 Competitive influences

Most companies within each industry or business sector do not operate in a vacuum and are therefore always seeking to react to moves by competitors. On many occasions this will involve the need to quickly respond with a change in the product offered to the market.

An example of this may be the need to accelerate the production or increase the quantity or the quality of a new item to the market, to ensure that it is in front of the launch of a competitor's product.

1.1.4.6 Risks inherent in the construction process

There are certain features of construction projects that create particular challenges and risks for clients, contractors and subcontractors. In addition, there are many variables that can be impossible to predict accurately in advance, yet have can have enormous impact on the ultimate cost and duration of the project.

Some of the features and variables which create risks in the construction process are:

- The complex nature of building works, for example:
    - entirely new build;
    - demolition of an existing building and its full reconstruction;

  - partial demolition and rebuild; or
  - refurbishment and/or extension of an existing building or structure.

- Unforeseen circumstances, for example:
  - unexpected ground conditions;
  - unpredicted weather conditions;
  - a shortage of resources;
  - a new design that proves impossible or very difficult to construct.
- The length of the contract (projects vary in the time required for completion from days to years).
- The number of participants (the number of parties involved in a project can have a large effect upon the contractual chain).

As can be seen from the above, many risks may be faced and may therefore need to be taken into account in considering the appropriate procurement strategy.

A major point to understand is that risks can be accepted by the client, can be transferred to another party by the client, or can be reduced or mitigated; but risks cannot and should not be ignored. The contract should address known risks head on and not leave the parties guessing at a later date.

If an identified risk can be removed altogether, so much the better, but this is unlikely to be possible in most cases. Elimination of risks will usually only occur during the course of the project, and so is unlikely to be possible at the commencement when the procurement strategy is being considered. However, risks can be managed in such a way so that the effect of the risks is greatly reduced.

It is often considered that risks should be held by the party best able to deal with that risk, and this is a factor that should be taken into account when considering the appropriate procurement strategy.

Clients may choose to allocate greater risk to themselves, thereby deriving potential benefits in terms of cost, time and quality. Alternatively, they may choose to attempt to divest themselves of risk almost entirely, with the resultant opposite effect, and with the added factor that the client may lose control of that risk item altogether.

Obviously, any risks that are transferred from the client to the contractor normally have attached to them a cost and/or programme impact, and this is in effect a risk 'premium', being a premium that (when recognised by the client) the client may rather not face.

The client's attitudes and drivers may change over time, as various risks begin to materialise. This is a factor that must be considered in the choice of a procurement route during the initial strategy meeting.

By way of analysis, it is usual to place the list of risks in various categories, depending on the nature of the risk involved and the potential ownership of those risks. Categories of risks could include some or all of the following:

- strategic risks, such as failure to obtain planning permission, or client funding problems;
- external risks, such as changes in the environment;
- project risks, such as overspends or delays to the programme;
- discovery risks, such as poor ground conditions or the like.

The listed risks are normally placed into a risk register, a contractual document that both lists the risks and provides a description of the impact that the risk in question would have on the project, should that risk event occur.

The overall list of risks can be as long or as short as is necessary (and this may well depend upon the complexity of the project). These risks will either be generic (common to most if not all construction projects) or will be specific to the particular construction project in question.

Generally, those risks that are external to the project are less able to be controlled in the timing or extent of impact upon the construction project, whilst those risks that are internal to the construction project can be more easily managed and can be allocated to the appropriate party in the construction project.

In the analysis of the risks in the risk register, a risk scoring exercise is normally undertaken. Each risk event is scored by reference to a pre-agreed scale of the likelihood of the event occurring. An element of judgment is inevitably needed as the project team may not necessarily be aware of the likelihood of client risk events occurring.

To avoid bias in the results, often the scoring is undertaken by various different disciplines in the client's procurement team and the scores are then averaged for each risk event. The risk events are also scored in a similar way on the basis of the impact that the event in question would have on the programme should the event occur. Consequently, it may be that some risk events are more likely to occur but would have little impact upon the project should they occur and vice versa.

This resultant risk profile may make it plain that the originally envisaged procurement strategy needs to be changed. The opportunity to deliver real benefits to the client (and for these to be accurately encapsulated within the contractual structure) is obviously best obtained if decisions to change the procurement strategy are made as early as possible, given that it is a well-known fact that any 'window of opportunity' to gain benefits progressively closes as the project continues.

In addition, this risk profile coupled with the client's attitude to risk (which may be an attitude developed through free choice, or, particularly in the case of public sector clients, may be an attitude dictated to it by statutory or other regulatory obligations) helps to decide upon the procurement route to follow to satisfy the procurement strategy, and which allows flexibility to deal with the anticipated risks that may be faced.

A chosen procurement route should be made with a clear awareness of the likelihood of occurrence and severity of the impact of risks. If a procurement route is chosen in this way, the project should be less severely impacted if the risk event in question actually occurs. Undertaking a risk analysis at this early juncture means that it can be updated throughout the construction process and can be used as a management tool to assist with the management of risks as they occur.

In some contract forms (the NEC forms, for example) the use of risk registers is an established way of recording risks during the construction process, and procedures are put into place to allow the parties to agree whether the risk should be carried by one party or the other, or whether the risk should be shared, or even if the risk can be eliminated altogether.

This approach is particularly useful if the allocation of the risks considered at the procurement stage are dealt with on a similar basis, and if used correctly can be utilised to anticipate and minimise the impact of potential risk items in the future.

1.2
KEY DRIVERS IMPACTING ON THE VARIOUS PROCUREMENT ROUTES

Armed with the procurement strategy as outlined above which has been developed from the client's attitudes and drivers in respect of cost, quality and time and with the risk profile developed from an analysis of the anticipated risks, the appropriate procurement route can be considered.

Obviously the transition from a concept to an actual project requires a formal agreement (a contract) which sets out who does what and for how much, how it should be done, and over what period, and allocates the risk between the parties.

Like all contracts, construction and engineering contracts are largely about the prior allocation of risk in addition to setting out the obligations and liabilities of the parties. When considering the allocation of risk, the following points may need to be considered:

- Is a contract with the lowest possible overall risk to the client required?
- Is there a restriction (legal, statutory, political or physical) as to which risks the client can accept?
- Particularly in the public sector, what are the priorities in apportioning the risks concerning time, cost and quality, and where are the speculative risks intended to lie?

With these points in mind, the procurement route used should deliver the procurement strategy that it has been decided should be used, or which, in some cases, it is required or stipulated must be used. In respect of construction projects in the private sector the choice of procurement strategy/procurement route is generally unregulated, in the UK at least.

As with the issue of the procurement strategy, the question of the procurement *route* is often best considered under the headings of cost, quality and time.

1.2.1 Cost

The projected value of the construction project will almost certainly have an effect on the procurement route chosen. For example, you are unlikely to decide upon a procurement route that includes burdensome administrative procedures for a construction project with a very low value.

Another consideration would be the appropriate valuation option. There are a number of different valuation options available, which are reflected in the various types of procurement routes.

These contractual options include:

- lump sum
- remeasurement
- activity led, target cost
- cost reimbursable

These options range from the lowest risk to the client, but the highest risk to the contractor (which is the lump sum method); to the highest risk to the client, but the lowest risk to the contractor (which is the cost reimbursable method).

Following on from this is the question of how payment is made, as the selection of an appropriate method of payment for the construction project will have implications on the risk profile of securing the project, from the standpoint of both the client and the contractor.

A commitment by a client to make prompt payments at the head of a supply chain fosters greater confidence in the ready flow of funds throughout the supply chain and serves to reduce payment risks at all levels. Secure and guaranteed cash flow is the 'life blood' of any business and contracting firms are no different in that respect. Accordingly, the selection of the way in which payments are to be made under the contract has a considerable impact on the way in which the client and the contractor will view the risk profile at tender stage.

In respect of public sector clients in the UK at least, the Government has produced a number of fair payment guidelines and practices as well as charters setting out what it considers to be fair payment terms for public sector works.

Consideration may also need to be given to BS 8903: 2010 and any further iterations thereof, which is a British Standard which gives recommendations and guidance on how to adopt and embed sustainable procurement principles and practices across an organisation and its respective supply chains. The principles set out in BS 8903 are applicable to both public and private sector organisations.

There are also a number of different payment options available to the project team, which are reflected in the various types of procurement routes.

These options include:

1.2.1.1 Single lump sum (at the conclusion of the project)

This approach is only likely to be used for the smallest value projects and those of the shortest duration. If the client demands such a payment method, then this introduces the highest level of risk for the contractor and s/he is likely to either decline to accept the project award or add a cost premium to the project to seek to cover his/her higher risk profile.

This payment method has, on the other hand, the lowest risk profile for the client in that it is not required to make any outlay until the project is complete. It has the option to refuse to make any payment to the contractor until the whole of the project is completed to his or her satisfaction, a situation which could be abused by the client to the greater risk of the contractor.

1.2.1.3 Stage payments

Many procurement routes require the contractor to be paid only when a stage or portion of the project is complete, whether or not this is verified by a third party. Unlike the lump sum method noted above, this greatly assists the cash flow of the contractor in that s/he receives payment for the work executed progressively throughout the life of the project.

This serves to reduce the contractor's risk exposure and allows a greater confidence when initially submitting his/her tender price for the project. However, contractors may have some concern over this payment method, as there is the possibility that the 'signing off' of a stage payment could be withheld by the client's representatives until such time as the whole of the works within the relevant stage is complete. As before, this has the potential to be open to abuse and materially change the contractor's risk exposure for the project. In England & Wales, at least, this risk is somewhat mitigated given that judge-led law has clarified unequivocally since February 2015 that an employer can only pay less at interim stage where requisite notices envisaged by statute have been issued in the correct form and on time.

1.2.1.4 Periodic payments

The traditional payment method is to undertake regular periodic payments (usually monthly) based upon the quantum of work undertaken during the period in question. This serves to have a lower risk profile for the contractor in that he/she is assured of a regular flow of funds, based upon the amount of work actually carried out.

From the client's perspective, this method of payment has the potential to be more costly, in that it requires a greater input by various members of the project team to undertake the interim valuations. It could also be the case that this approach introduces a greater risk of disputes arising over disagreements about the quantum of work undertaken.

1.2.2 Quality

The required quality and standards will naturally have an effect on the procurement route chosen. For example, if the client wishes to have complete control over the design during the construction phase, you are unlikely to select a 'design and build' option.

The choice of procurement options range from those where the client retains responsibility for and control of the design team and the design quality throughout the process (for example by following the 'traditional' procurement route) to those where the client may have a lack of control over design decisions with the result that the project may not be perfectly suited to his/her needs when completed (for example by following a 'turnkey' approach).

Other issues that may arise relate to matters such as:

- how 'buildable' the design may be;
- the potential for over-design and/or over-engineering;
- the opportunity to use the expertise of the contractor in the design process;
- the lack of flexibility available to the client to make changes during the build process; and the checks on quality that the client wishes to be able to make.

All of these issues need to be taken into account when deciding upon the procurement route to follow.

1.2.3 Time

The importance of time for commencement and time for completion are key factors in deciding upon the procurement route to choose. For example, if the client wishes to have an early start on site, you are more likely to choose a 'design and build' or 'management contracting' approach rather than a 'traditional' procurement approach.

The choice of procurement options range from those where the scheme needs to be more or less fully designed before tenders are sought, thus extending the pre-tender period (for example by following the 'traditional' procurement route) to those where minimal design information is available at the commencement of the project but works on site can still commence rapidly (for example by following a 'design and build' approach).

Other issues that may arise relate to issues of what information is available at the tender stage and the commencement on site stage; what programming input is required from the contractor, and when that input is expected; and what design input is required by the contractor.

1.2.4 Other factors

In addition to the risk profile referred to above, and the key issues of cost, quality and time, the choice of procurement route may also be influenced by the following:

1.2.4.1 The nature of the project

Is the project for a completely new detached building; an extension to an existing building; the refurbishment of an existing building; the restoration of a historic structure; or a repair and maintenance programme involving many buildings?

1.2.4.2 The complexity of the project

Is there something unusual about the size, complexity or location of the works; are there site problems of access, storage or movement; is it a simple project that only involves the basic trades and skills of the industry; does an innovative design demand sophisticated construction methods; is there specialist subcontractor's work with a design content; is there a high content of specialist engineering installations; is this a single construction operation, phased work, or part of a term programme?

1.2.4.3 Environmental issues

There may be a need to change production methods to take into account environmental policy or sustainability issues.

1.2.4.4 The anticipated involvement of the client

Should the design be wholly in the hands of the client's consultants; can some detail design be placed as a contractor's responsibility; should there be provision for design by specialist subcontractors; to what extent does the client wish to control selection of specialist subcontractors; what measure of control will the client wish to exert over materials and workmanship; how much reliance can be placed on performance specified requirements?

1.2.4.5 The need or desire to build relationships with the supply chain

Is a long-term relationship with a supplier or the supply chain required so as to provide continuous improvement?

1.2.4.6 The anticipated need for changes or variations

Is there a likelihood of design changes during the course of the works; can the contract satisfactorily accommodate variations and the valuing of such work; to what extent might approximate quantities or provisional sums be required?

1.2.4.7 The anticipated involvement of the client

Should the design be wholly in the hands of the client's consultants; can some detail design be placed as a contractor's responsibility; should there be provision for design by specialist subcontractors; to what extent does the client wish to control selection of specialist subcontractors; what measure of control will the client wish to exert over materials and workmanship; how much involvement will the client wish to have in the design and construction process; how much reliance can be placed on performance specified requirements; the extent to which the client may wish to change his/her mind or alter the specification during the course of the works.

1.2.4.8 The possible need for collateral warranties (or the granting of third party rights)

On an increasing number of projects (particularly in the commercial world) there will be the need for the client to either introduce a requirement for him/her to receive the benefit of a warranty from another or to be required to deliver a warranty him/herself to others. A warranty is given or received where it is necessary, or desirable, to introduce a direct contractual link between two parties that are not directly in contract with each other: say, the employer and a designing mechanical services subcontractor.

As the preparation and execution of multiple warranties on each project can be a cumbersome and time consuming process, there is an alternative option of merely granting third party rights to others, which seeks to effect the same end result. This has been available since the enactment of the *Contracts Rights of Third Parties Act* 1999, although there is evidence that the 'take-up' of this option has been slow. It appears to have been treated with some suspicion and the construction industry appears to prefer the tried and tested collateral warranty approach given that notice often needs to be effected independently before third party rights can be called upon. However, third party rights in contracts are now more institutionally acceptable, and in growing in use by many major developers.

1.2.4.9 The desirability of performance protection for the project

There is always the risk present that the contractor will fail to honour his/her contractual obligations, most seriously due to insolvency. It is therefore possible for clients to seek to protect their financial position in the event of non- performance, by the taking out (at project commencement) of financial products such as bonds.

There is a wide variety of different types of bonds on the market, the more usual ones being payable upon the contractor's non-performance and/or insolvency. The bonds may be unconditional 'on demand' bonds or, more commonly, conditional bonds which rely on a failure or breach by the contractor to trigger release of the bond. Each form has its own risk attached and specific advice should be sought on their use and consequences.

The bonds taken out should be considered as a form of insurance policy, because if no claim is made, the amount paid for the bond is non-returnable. Many corporate clients will have specific company rules which govern the situations in which a bond is to be procured for a project and clients invariably rely on credit agency data to guide them in this area. The bond quotation figure that the contractor obtains is also a very good indicator of his/her financial standing, as the price quoted to him/her by the bondsman will be directly related to his/her perception of the likelihood of failure.

For these reasons, many contractors will seek to offer clients alternative means of protection, in the form of a parent company guarantee (PCG), although these instruments should be treated with some caution, given that they might be considered of limited value and protection (particularly if the parent company's own financial standing is unstable). Many standard forms provide for the procurement of such security.

1.2.4.10 The possible need for termination of the contractor's employment

There is always the risk that the client will need to dispense with the contractor's services prior to the completion of the project, and this will usually be because of a lack of performance or inadequate quality of work or the failure of the contractor to achieve defined time criteria. There is also the possibility that the client may be forced to terminate due to external factors, such as market conditions within his/her own business and industry.

The client will usually have very strong (and sometimes automatic) rights of termination in the event of the contractor's insolvency; as will the contractor if the roles are reversed.

It is important that the client has the flexibility to be able to terminate should s/he choose to, or should circumstances be forced upon him/her, and it is normally required that the resultant financial adjustments made as a result of the termination action are a fair and equitable balance between the parties.

1.2.4.10 The desirability of a sensible dispute resolution process being in place

There has been acknowledgment in recent years that the traditional and historic means of dispute resolution were becoming increasingly unacceptable to all parties due to cost and excessive time taken. This had served to provide a strong disincentive to seeking redress for alleged defaults. Arbitration, which is still used extensively internationally, has declined in use domestically in the UK as a primary form of alternative dispute resolution and most recent contracts, for example, now state that litigation, rather than arbitration, is the default 'final' dispute resolution process.

Adjudication, introduced under the *Housing Grants, Construction and Regeneration Act* 1996, is now considered to be the principal dispute resolution process for the construction industry in the UK. Whilst adjudication has its critics, and it has been slightly amended by the *Local Democracy, Economic Development and Construction Act* 2009, it is still considered to be a relatively quick and inexpensive, and reasonably effective dispute resolution process that produces a 'temporarily binding' decision which then is enforced in the High Court Queen's Bench Division (specifically, the Technology & Construction Court).

In more recent years, mediation (or similar processes) are growing in use in the construction industry and certain forms write the process in as a precursor to other ADR or more formal methods. Nevertheless, parties are often not obliged to refer disputes to mediation and the outcome is not, as a matter of course, binding on the parties. Mediation does however have many benefits and a track record of success but those issues are beyond the scope of this reference guide. Adjudication does have the benefit of being a statutory requirement, in a commercial setting, provided certain hurdles are overcome.

It is important that clients have the flexibility to be able to decide which dispute resolution method is most appropriate for them on any particular project, given the particular circumstances and being mindful of the statutory framework prevailing at the time.

1.2.4.11 The need for standard form contract amendments

Clients may well have the need to make amendments to the published text of standard form contracts because of corporate requirements, regulatory or statutory considerations, internal rules or processes or other contractual obligations which they are required to comply with.

It is also acknowledged that clients will make amendments (sometimes heavily) to standard forms if they (or their advisers) perceive that the published contract favours the contractor and wish to rebalance that risk. Unfortunately, on occasion, this reallocation goes too far and the contractor is usually forced to price for the enhanced risk, with resultant higher tender figures.

In any event, amendments should be made with caution, and where possible, with expert advice. Intelligently amended standard terms can add clarity and substance to a contract, but poorly amended contracts lead to contractual chaos and conflicting and/or incomplete provisions create huge uncertainty and are the source of many disputes.

These factors must be remembered as secondary considerations when considering the choice of the procurement route and the form of contract. All of the above factors are normally taken into consideration before choosing a procurement route.

In summary, before you can think about which contract form is most appropriate, you will need to:

- Ascertain client drivers and replay them back to ensure they are understood by the project team;
- Consider providing options for achieving client outputs based on certainty of time, cost and quality;
- Think further about various factors which may impact on risk – some you will be able to eliminate in some way, some you will not. You will be able to manage these via:
    - risk registers;

    - risk analysis at the outset and subsequently during the project; and
    - appointing the right team, which is as much about previous track records and styles of work together as it is about having suitable qualifications.

- The procurement route will depend upon the client drivers and risks identified as well as the impact of socio-economic factors and the legislative environment.

1.3 PRINCIPAL PROCUREMENT ROUTES

Before we can choose the form of contract, we first need to recap on the principle procurement routes available and the advantages and disadvantages of each.

1.3.1 The 'traditional' approach

The traditional approach is sometimes also referred to as conventional procurement.

This is an approach that became standard practice in the building industry in the 19th century following the emergence of the general contracting firm and the establishment of independent client consultants. The main feature of this approach is that the design process is kept separate from the construction process.

Consequently, for this method to be truly effective, full documentation (including the design) ideally needs to be in place before the contractor can be invited to tender for carrying out the work.

Unfortunately, this objective is not always achieved, and as a result the effectiveness of this procurement route is lessened and the project can then run into the problem of being designed as it is being built rather than being designed before it is built.

Using this traditional procurement approach, the client contracts with a project manager/quantity surveyor, architect or an engineer to carry out the design; the client also enters into a separate contract with a contractor who carries out the building works.

The appointed architect, engineer (or a project manager or surveyor, if appointed) will normally also fill the role of the contract administrator.

The contract administrator then supervises the construction of that design to ensure the works are completed economically and efficiently.

However, when issuing certificates under the contract or when undertaking decision-making under the contract, a contract administrator is required to act independently, impartially, honestly and fairly between the client and the contractor. There is an element of inherent duality here, which will be explained later on, which affects the powers and obligations of that third party certifier, albeit the way in which each standard form deals with it can be slightly different.

The architect or the engineer (or the project manager, if appointed) does not enter into a direct contract with the contractor but will have its own appointment with the client.

In carrying out the required work, the contractor normally directly employs both subcontractors and suppliers of services, goods and equipment.

Out of the relationships between the various parties noted above, the issue of privity of contract (the basic rule that only parties to a contract can enforce the contract) is particularly relevant. Therefore, there is no direct contractual link between the client and a contractor's subcontractor (for example) but the use and spread of collateral warranties, discussed above, has resolved some of the associated problems.

Under the traditional approach of procurement, the contract price (normally a lump sum price, subject to adjustment only as a result of variations) is often based upon a Bill of Quantities provided by the client which quantifies, as far as possible, every aspect of the works. The measurement of the quantities is normally based upon the *Standard Method of Measurement for Building Works*, or similar, published by RICS.

However, as an alternative to a Bill of Quantities being produced, contractors can be required to price the works on the basis of a set of drawings and specification of the works being provided. Also, it is quite common for remeasurement contracts, target cost contracts and cost plus contracts to be used as a means of pricing the works under the traditional procurement approach.

The traditional approach to procurement remains popular with many participants in the construction industry, and this could be because it is very familiar to most client and contractors.

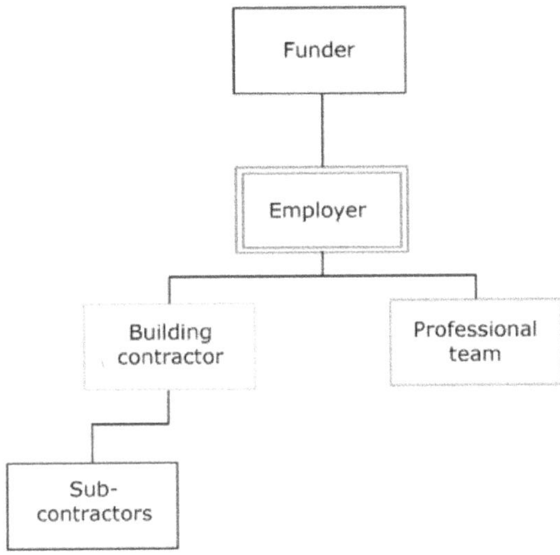

Advantages

Some of the advantages of the 'traditional' procurement approach are:

- familiarity among contractors and consultants – the roles and responsibilities are well understood;
- the client retains responsibility for and control of design team;
- there is direct reporting by the design team to the client to ensure that quality control is maintained;
- the client has an independent professional in the role of contract administrator monitoring the project;

- there is certainty of price (if the work is fully designed in advance);
- priced Bills of Quantities provide a basis for variations to be priced at tendered rates;
- as all prices are based on the same information:
    - there is no need for a contractor to build in a risk premium;
    - it is easier to analyse the prices;
    - the lowest price is usually the best value for money;
        - where elements of the building are not fully designed, provisional sums may be used to allow for the later design of those elements.

Disadvantages

Some of the disadvantages of the 'traditional' procurement approach are:

- to be effective, it requires the scheme to be more or less fully designed before tenders are sought – this often results in an extended pre-tender period;
- the fragmented design and construction process and responsibility can lead to disputes, for example in respect of whether construction defects are really design defects or whether they are construction defects;
- there is the potential for over-design and/or over-engineering;
- the contractor is not involved in the design process and therefore is not required to 'buy in' to the design;

- the client retains responsibility for the design team performance;
- a fixed lump sum price is rarely actually achievable;
- a contractor may price the work to win the job rather than providing a price that properly reflects the work to be carried out. This can encourage a claims culture if the submitted price was too low because of market forces;
- the use of provisional sums and the power of the architect or engineer (or project manager, if appointed) to issue instructions for additional or varied works can lead to price escalation.

Summary

In respect of cost, whilst there may be certainty of price (if the work is fully designed in advance) and priced Bills of Quantities provide a basis for variations to be priced at tendered rates, it is sometimes the case that a contractor may price the work to win the job rather than providing a price that properly reflects the work to be carried out and this can encourage a claims culture if the submitted price was too low because of market forces.

With regard to quality, whilst the client retains responsibility for and control of design team and there is direct reporting by the design team to the client to ensure that quality control is maintained, there is the potential for over-design and/or over-engineering and the contractor has little input into the design process.

In respect of time, to be effective, the 'traditional' approach requires the scheme to be more or less fully designed before tenders are sought and this often results in an extended pre-tender period. If the design is not completed before tenders are sought this can lead to the 'design as being built' approach being adopted, which invariably delays the on-site programme period.

1.3.2 The 'design and build' approach

The 'design and build' procurement approach is used where the contractor is to be responsible for undertaking both the design and the construction of the required work in return for a lump sum price.

There are variants on this option depending on the degree to which initial design is included in the client's requirements.

These variants can be grouped under three main categories:

- *Package deal or turnkey contract* (dealt with in more detail below).

  This is where the client settles on a complete package for design and build, often to some standard specification from a commercial firm.

- *Design and build contracts.*

  This is where the contractor is (normally) responsible for the complete design and construction of the works in question.

- *Contractor's design for specific elements only.*

  Strictly speaking, these are not design and build contracts, but are actually traditional 'work and materials' contracts which include for limited design provision relating to an identified portion of the work.

The main feature of the design and build procurement approach is that the design and the construction process are both carried out by the contractor.

To ensure that the client obtains what he/she is seeking in respect of a finished building it is essential that the client specifies exactly what he/she requires and checks that this is matched by way of what the contractor offers to provide.

Therefore, although it is not necessary for full documentation (including the design) to be in place before the contractor can be invited to tender for carrying out the work, it is important that the client's brief and his/her requirements are clearly set out.

Using the design and build procurement method, the client contracts with an agent (usually an architect, a quantity surveyor or a project manager) to administer the contract on its behalf. The client also enters into a separate contract with a contractor for it to carry out the design and construction of the building works.

The agent does not enter into a direct contract with the contractor. In carrying out the work, the contractor normally employs both subcontractors and suppliers of services, goods and equipment.

Privity of contract, as discussed above, is particularly relevant. Therefore collateral warranties are commonly used in design and build to provide a direct contractual link between parties that require such a direct contractual link, for example, between the contractor, subcontractors and a funder.

Under the design and build approach of procurement, the contract price (normally a lump sum price, subject to adjustment only as a result of variations) is usually the price for the works provided by a contractor and this is most often based on a set of drawings and a specification of the works to be provided.

The design and build procurement approach is popular as the liability for both the design and the build is with the contractor, and there is less chance of a liability of a defect 'falling between the two stools' of design and construction.

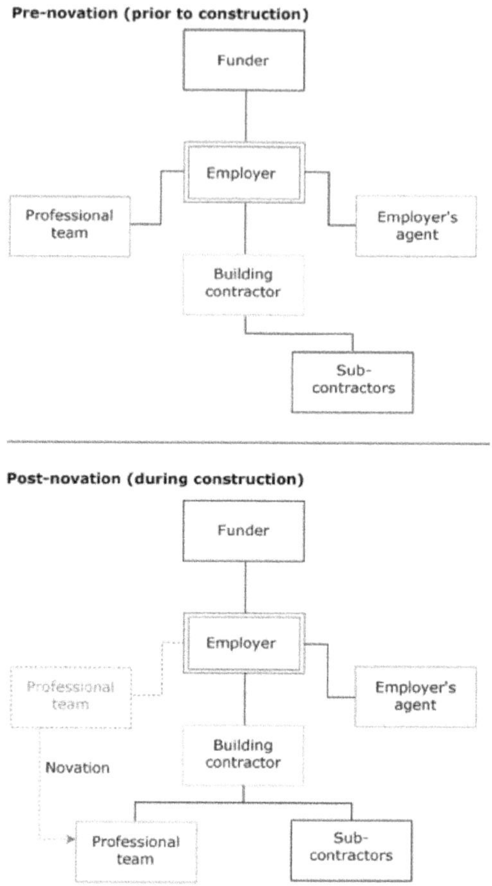

Advantages

Some of the advantages of the 'design and build' procurement approach are:

- *Speed of delivery from concept to completed building.*
  There is normally a much quicker delivery time than for traditional procurement. In its simplest form, design and build allows work on site to begin earlier (that is before the design is fully complete) than under traditional forms of contract, because of the level of design control given to the contractor. Normally, the design and build procurement approach allows

programmes and budgets to be more easily met and the speed of construction is also often quicker;

- *Single point responsibility.*
  The contractor is responsible for the design and the construction. Therefore the client should have a single point of responsibility and liability against the contractor. This is more advantageous than the traditional forms of contract where the client has entered into separate construction and design agreements. A common problem with the latter approach being that if a claim is made, the contractor, architect or other design consultants may argue over the extent of their own individual responsibilities. Consequently, there is less likelihood of claims being made by the contractor in respect of the split responsibility of design, and the contractor will be unable to make a claim for late design instructions being issued (other than for changes subsequently made by the client);

- *Acceptance of design.*
  Because the contractor is responsible for the design and the construction, the contractor and his/her supply chain are involved in the production of the design to be used, and hence 'buy in' to that design. Also, it follows that the design is more likely to be 'buildable' than may be the case under other procurement methods;

- *Novation of design.*
  There is normally the facility for the client's own designers to be novated to the contractor. This approach has several perceived benefits, including; the client may have used those designers many times previously and will be happy with the quality of their work; the design team is likely to be more attuned to the client's requirements; the design team can continue with the contractor where they left off with the client; and some clients believe that through the novation of their own designers they (in effect) have an independent voice in respect of the contractor's subsequent design intentions;

- *Cost certainty.*

  It is generally the case that, as the contractor can use his experience and expertise in providing a design that allows him to buy goods and services which allows him to obtain the best buying margins, the design and build procurement route can be more cost effective and can provide more cost certainty – provided, of course, that the client does not continually change the brief;

- *Less client management/consultant involvement required post contract.*

  This therefore results in lower management costs and lower consultants' fees;

- *Unless stated otherwise, there is a duty of fitness for purposes on a design and build contractor.*

  This is more onerous than the normal duty of 'reasonable skill and care' imposed on a design consultant. Of course, often the contract does state otherwise, and the contractor's design obligation is limited to a duty of reasonable skill and care. One of the reasons for this is that most professional indemnity insurance policies do not cover for a fitness for purpose obligation.

Disadvantages

Some of the disadvantages of the 'design and build' procurement approach are as follows:

- *Price.*

  Initial price may be higher as the contractor may build into his price a 'risk premium';

- *Ongoing expense.*

  Post-contract variations can be more expensive, and it is often more difficult to monitor the additional charges raised (particularly where works are priced on the basis of a specification and drawings, for example);

- *Less control.*
  The client has less control and influence over design matters;

- *Change limited.*
  There is only limited scope for the client to make changes to his requirements once the client's requirements and contractor's proposals have been agreed, otherwise the cost consequences may be prohibitive. If the client does not have a firm and robust set of client's requirements he may be given a design that he did not want, or may be required to pay considerably more to obtain the design that he did require;

- *Discrepancies between documents.*
  There may be a conflict between the client's requirements and the contractor's proposals unless both documents are carefully checked. This conflict can be obviated by making it clear in the contract which document takes precedence;

- *Design quality.*
  It is often perceived that the contractor is driven by price rather than by design standards and, therefore, it is often considered that the design and build procurement route is not the appropriate route to use where a high quality design is required, unless a robust specification is included within the client's requirements;

- *Dual loyalties.*
  There is the potential for dual loyalties of a novated architect. Also, there could be complex legal issues in relation to the novation process. Contractors need to ensure that they are able to pursue claims against the novated consultants for any pre-novation losses they may suffer as a consequence of breaches of duty by the consultants pre-novation. The question of the quality achieved can be an issue because of the lack of control that the client has over the architect. The architect acts for the contractor not for the client;

- *Perceived lack of independence of the architect.*

  If the client wishes to take independent advice on design issues following the building contract being entered into, or the novation of the design team to the contractor, the client will have to pay additional fees.

In respect of Cost considerations, the initial price may be higher as the contractor may build into his price a 'risk premium', and post-contract variations can be more expensive. As there is a single point responsibility between design and build, there is however less likelihood of claims being made by the contractor in respect of the relationship between the design function and the build function. The process can be more cost effective because the contractor can use his experience and expertise in providing a design that allows him to buy goods and services which allows him to obtain the best buying margins.

With regard to Quality, the design is more likely to be 'buildable' than may be the case under other procurement methods. However, the client has less control and influence over design matters, and there is a degree of inflexibility in respect of design changes to be made. Also, because the architect acts for the contractor and not for the client, the check on quality at that level is not available to the client.

With respect to Time, the process allows for speed of delivery from concept to completed building, and there is normally a much quicker commencement on site and delivery time than for traditional procurement. This process allows programmes to be more easily met and the speed of construction is also often quicker.

1.3.3 The 'management contracting' approach

This involves a contractor managing the carrying out of the work through works contractors, and those works contractors being contractually accountable to it.

The client normally engages the management contractor to take an active role in the project at an early stage, and, because of this, the management contractor is normally an experienced contractor.

The overall design is the responsibility of the client's consultants, but the management contractor is normally responsible for defining packages of work and then for managing the carrying out of those work packages through separate trades or works contracts.

The management contractor is not employed to undertake the work but is employed to manage the process. All the work is subcontracted to works contractors who are directly employed by the management contractor. The client usually needs to be given the opportunity to approve the terms and conditions of the trades or works contracts before the packages are subcontracted.

The management contract will usually include both a pre-construction phase and the construction phase. Documentation will start with project drawings, a project specification and a cost plan. This information will then be formed into documents on which competitive tenders can be obtained for the work packages.

The management contractor is responsible for the administration and operation of the works contractors. However, the management contractor is not liable for the consequences of any default by a works contractor so long as the management contractor has complied with the particular requirements of the management contract.

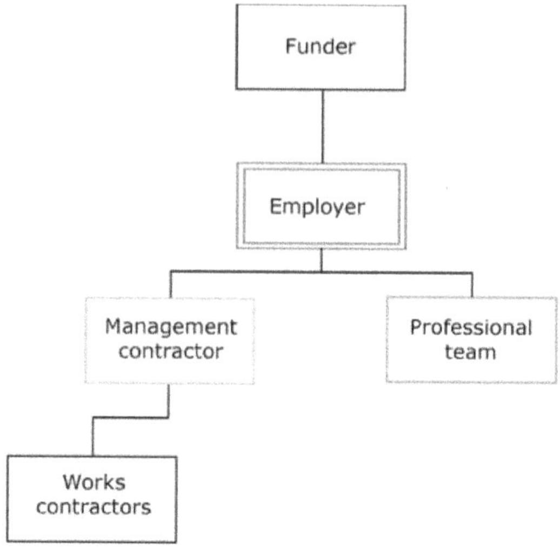

Advantages

Some of the advantages of the 'management contracting' procurement approach are as follows:

- *Fast start.*
  It is particularly beneficial for fast-track complex projects where minimal design information is available at the start of the project;

- *Strong feasibility and programming interface.*
  It allows for early 'buildability' and programming input from the management contractor acting as a consultant;

- *Single point management.*
  There is a single point contractual and payment arrangement for the client with the management contractor (rather than to all of the individual works contractors);

- *Price.*
  The preliminaries and management fee can be fixed, therefore allowing for a degree of certainty on price;

- *Quality*,
  which can be controlled by the design team;

- *Flexibility.*
  There is great scope for client changes.

Disadvantages

Some of the disadvantages of the 'management contracting' procurement approach are as follows:

- *Higher risk for client.*
  This procurement approach is a low risk strategy for the management contractor as he/she has little responsibility for package contractor defaults, bankruptcy, etc.;

- *Prime cost in nature.*
  Although guaranteed maximum price can be achieved, the process is still fundamentally prime cost in its nature – which is an approach that many contractors naturally prefer;

- *Scope for substantial cost increases.*
  There is often a tendency for the initial cost plan to be adjusted upwards.

1.3.4 The 'construction management' approach

The construction management procurement approach is a method whereby the construction manager (often a contractor or someone with sound project manager skills) undertakes to manage the carrying out of the work through trade contractors. Although the trade contracts are arranged and administered by the construction manager, contractually they are the client's risk, and the client forms direct contractual relationships with the trade contractors.

Obviously for this to be effective, the construction management appointment and the trade contract for each trade contractor must be compatible.

The construction manager is paid a fee as though he were any other consultant to cover for his/her staff costs, and overheads, etc. As part of his/her duty, the construction manager is required to co-ordinate the trade contractors.

The construction management procurement approach is a method where overall design is the responsibility of the client's consultants.

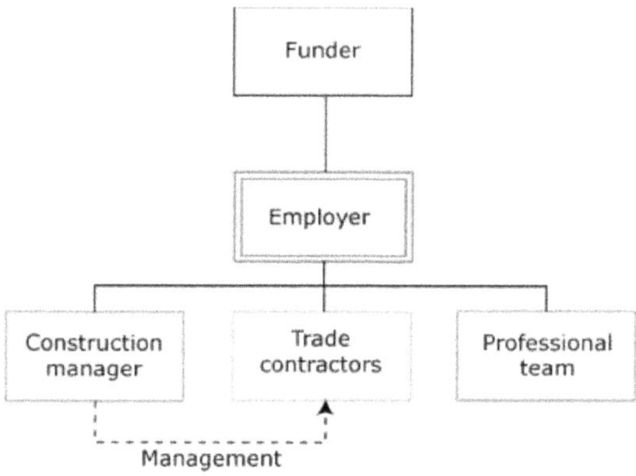

Advantages

Some of the advantages of the 'construction management' procurement approach are as follows:

- *Enhanced client control.*
  The client is more in control of the process through the employment of the construction manager and the direct contracts with the trade contractors. The client has better cost and budgetary control, and there is a higher degree of control due to the project being broken down into trade packages;

- *Integrated design.*
  There is a fully integrated design and construction process;

- *Aligned interests.*
  The construction manager acts on the client's behalf whereas a traditional contractor primarily acts in its own interests;

- *Flexibility.*
  Changes to suit the client's requirements are relatively easy to accommodate;

- *Helps avoid disputes.*
  There is a reduced potential for claims.

Disadvantages

Some of the disadvantages of the 'construction management' procurement approach are as follows:

- *Size required.*
  The project needs to be sufficiently large or complex to be cost effective;

- *Administration by client.*
  The process requires the client to have mechanisms for entering into direct contracts with trade contractors and for making monthly payments to many individual contractors;

- *Internal expertise required.*
  The client needs to have the necessary experience and internal management ability to operate this procurement process;

- *Higher client risk.*
  The client retains the contract risk of non-performance of the trade contractors and it retains responsibility for the design team performance.

In respect of Cost, the client is more in control of the process through the employment of the construction manager and the direct contracts with the trade contractors and therefore has better cost and budgetary control. However, the client retains the contract risk of non-performance of the trade contractors, and the process requires the client to have mechanisms for entering into direct contracts with trade contractors and for making monthly payments to many individual contractors.

With regard to Quality, integration of the design process with the construction process means that innovative and practical alternatives can be incorporated into the building more easily, and the client retains control of the design team performance.

With respect to Time, because of the integration of the design process with the construction process the lead-in period before commencement can be shortened, although care needs to be taken that because the process is very flexible, the on-site period may be prolonged because of indecision by the client.

1.3.5 The 'turnkey' approach

A turnkey contract is a business arrangement whereby a project is delivered in a completed state.

Therefore rather than contracting with various parties to develop a project in stages, a client enters into a contract with one party (normally a developer or a contractor) to finish the entire project without any further input from the client.

The developer or contractor is separate from the client, and the project is handed over only once it is fully operational. In effect, the developer or contractor is finishing the project and 'turning the key' over to the client.

This type of arrangement can be used for construction projects ranging from single buildings to large-scale developments.

Under a traditional lump-sum contract, the client agrees to pay the contractor to complete a project that is built to the contractor's specification. The client is given many opportunities to make decisions throughout the project, and to make changes as needed.

In a turnkey contract, the client is generally left out of the building process entirely as the developer or contractor handles all decisions and problems related to construction.

Advantages

Some of the advantages of the 'turnkey' procurement approach are as follows:

- *High motivation to complete quickly.*
  The developer or contractor normally owns the project until it is complete, and therefore it has the financial motivation to complete the job as quickly and efficiently as possible;

- *Finance friendly.*
  A turnkey contract provides more time for a client to seek financing and investors before s/he is required to pay for a completed project;

- *Expert engagement for the uninitiated.*
  An inexperienced client does not need to make difficult construction decisions, as those decisions are left entirely in the hands of the developer or contractor.

Disadvantages

Some of the disadvantages of the 'turnkey' procurement approach are as follows:

- *Lack of control.*
  The client maintains over design and construction decisions, which may mean that the project is not perfectly suited to their needs once it is complete;

- *Conflicting priorities.*

  In the case where a developer or a contractor has some financial long-term interest in the project, (for example, a developer or a contractor will construct a retail establishment for a client, and the developer or contractor may receive a percentage of the gross receipts for a specific period of time) this may encourage the developer or contractor to make construction decisions based on the long-term needs of the project, rather than just the short-term decisions needed to get the project completed.

In respect of Cost, the contractor/developer normally owns the project until it is complete, and therefore it has the financial motivation to complete the job as quickly and efficiently as possible. A turnkey contract provides more time for a client to seek financing and investors before he/she is required to pay for a completed project.

With regard to Quality, the client normally has a lack of control over design decisions, and this may mean that that the project is not perfectly suited to his/her needs when completed. If the contractor/developer has some financial long-term interest in the project this may encourage them to make construction decisions based on the long-term quality needs and life-cycle cost interest of the project.

With respect to Time, the process can take more time to set up initially, but as the contractor/developer has a vested interest in achieving early completion (because often its payment is linked to completion) the period on site can be shortened.

1.3.6 The 'partnering' approach

Partnering is a concept that can be applied to many other procurement routes and is often a co-operative relationship between business partners formed in order to improve performance in the delivery of projects.

Partnering is best considered as a set of collaborative processes which emphasise the importance of common goals and raise such questions as how such goals are agreed upon, at what level are they specified and how are they articulated. It is applied either in a project situation, known as project partnering, or in a long-term relationship known as strategic partnering.

There are two views in the construction industry regarding the role of the contract between the client and the contractor in the partnering process:

(1) The first view is that partnering is all about co-operation, dispute avoidance and self-improvement and that, as such, a successful project partnering agreement can be implemented independently of the contract, even when the contract contains clauses that are not in alignment with the co-operative principles of partnering.

(2) The second (and alternative) view supports the use of the project contract to reinforce the elements of a partnering arrangement. This can be in the form of a traditional standard form contract amended to enforce a partnering agreement or in the form of partnering-type standard contract.

Partnering is normally operated in one of the following ways:

- a traditional construction contract with a separate partnering charter, which is often a series of behaviours to which parties agree to adhere, albeit on a non-binding basis (see the JCT Partnering Charter Non-Binding, 2011 edition, as an example); or

- a two-party contract aligned to partnering. Under this method, the construction contract chosen reflects the principles of the intended collaborative relationship between the parties. This remains a two-party approach and relies on contract conditions covering the partnering aspects. The crucial difference to the first approach is that the parties are contractually bound to working co-operatively; or

- a multi-party partnering contract. A major difficulty in the acceptance of such a multi-party contract is the perceived complex legal situation in regard to the responsibilities and liabilities in a multi-party situation.

In the partnering approach, negotiation rather than competitive tendering is the key. The risks that both parties face as a result of the project in question are made transparent at the commencement of the project, and the partners to the partnering approach share those risks on the basis of which party can best bear those risks and/or insure against those risks.

The essence of any partnering agreement now involves a duty of good faith, mutual co-operation and trust between all parties involved in the construction process.

Advantages

Some of the advantages of the 'partnering' approach are as follows:

- a reduction in the number of disputes;
- the benefit of early supply chain involvement;
- it is based on an open book and a win/win culture;
- there is integration of the design process with the construction process;
- the main benefits are generated from strategic partnering (multiple projects) rather than a single project.

Disadvantages

Some of the disadvantages of the 'partnering' approach are as follows:

- the partnering process can be abused by one of the parties;
- the partnering process requires more client resource to compensate for the less competitive environment, and the process can collapse when one party becomes disadvantaged;
- to be most effective, partnering needs to be practised and learnt over a series of projects and typically requires an early commitment in terms of management resources and direct costs;
- there are the direct costs of workshops, of training staff and of the more intensive early involvement of management in establishing the partnering approach.

In respect of Cost, because partnering is normally based upon an open book and a win/win culture there is an incentive for the contractor to pass on savings that can be found. However, the partnering process requires more client resources to compensate for the less competitive environment, and typically requires an early commitment in terms of management resources, workshops and training of staff.

With regard to Quality, integration of the design process with the construction process means that innovative and practical alternatives can be incorporated into the building more easily.

With respect to Time, because of the integration of the design process with the construction process the lead-in period before commencement can be shortened; although care needs to be taken that the on-site period is not prolonged because of unresolved design issues amongst the partnering team members.

In summary, you probably have a good understanding or refresher of the types of procurement routes typically available and their relative pros and cons. Having established what is of critical important to your client in terms of Time, Quality and Cost, you can then narrow down which of the many standard form contracts out there may fit your chosen route.

PART TWO

WHAT CONTRACTUAL OPTIONS DOES THE CLIENT HAVE TO ACHIEVE ITS AIMS?

2.1
THE CHOICE OF CONTRACT

The choice of contract available to meet specific needs can sometimes feel quite daunting. At times, this can result in parties sticking to what they know best when there may be other, more appropriate, options.

As has been explained in Part 1, the goal in effective procurement is invariably to achieve certainty of time (or completion), cost and quality. With these objectives in mind, there are a number of factors which may impact upon a decision to use a particular contract; namely:

- the various types and uses of the suites of contracts depending on the nature of a particular project;
- how design responsibility is apportioned and workmanship and change to the scope of the works managed to assure quality of the finish; and
- the possible payment structures available which will influence how a project is to be costed.

You are also urged to consult the extensive guidance published by each institution/body, to which this section is merely an introductory snapshot of some of the key considerations (both contractual and legal) when making a choice according to particular circumstances.

The nature, complexity and value of a particular project are all fundamental considerations when considering which contract to use from either the JCT, NEC or FIDIC (as well as PPC2000) suites.

The identity and roles of the parties operating the contractual terms are often the interface and way in which the contractor (or supplier) helps the employer (or client) achieve its objectives and so that is our starting point.

2.1.1 Appointment of the parties and supply chain relationship: cost and quality

This section considers more generally how the standard forms establish the relationships between the employer, the contractor and other members of the supply chain.

Aside from the obvious fact that an Employer is engaging a Contractor, how does the rest of the supply chain fit in?

The later JCT contracts (from 2005 onwards, culminating in 2016) place greater emphasis on integrated teams than previously where there was more of a separation between design and construction activities. Lawyers and other contract drafters have however been adapting JCT contracts for early contractor involvement for some time, engaging the contractor during the design phase to undertake preconstruction activities (The JCT then responded with its own form of Pre-construction Services Agreement).

This separation is also reflected in many of the FIDIC forms. NEC3 and PPC2000, by contrast, have more of a focus on early involvement of the supply chain in the project. The integrated approach can deliver benefits in respect of quality of design and increased control on costs, but requires greater resourcing, so will not always be suitable for small works.

With the rise in popularity of design and build as a procurement option over the past 20 years or so, transfer ('novation') of the employer's appointed design team to the contractor has become relatively common practice. The standard forms tend to have an outline novation provision but it is often specific amended to provide robustness to its terms. The exception is PPC2000, where in principle (given the multi-party structure) amendments for novation should not be necessary (members of the partnering team participate in the partnering contract by signing a Joining Agreement).

2.1.1.1 JCT (Supplemental Provisions for subcontractors)

Until the 2005 editions were published, JCT contracts included provisions for the nomination of specialist subcontractors and suppliers to allow the employer some control over choice and performance (see also PPC2000 and FIDIC). However, the insolvency and performance risks that employers assumed with nomination were unwelcome and an informal practice of 'specifying' subcontractors and suppliers developed. Still essentially a nomination process with similar pitfalls, this process and nomination became increasingly unpopular.

The 2005 edition and subsequent revisions replaced both processes with a 'named' subcontractor option (which had been in existence prior to 2005 in some forms) whereby the employer specifies in the Contract Particulars that the relevant Supplemental Provision (Part 1) applies. Thereafter the employer can step back from assuming the responsibility for selecting subcontractors, albeit there is a safe-guard in the form of a deemed Change if the Contractor needs to terminate the Named Sub-Contractor's employment for any valid reason and recover associated losses.

JCT produce a partnering agreement which has its merits. Faced however with its inherent history, it has somewhat lost out to others including NEC forms, FIDIC and PPC2000 in terms of integrated interfaces.

2.1.1.2 NEC3 (clause 10.1 – parties must have trust and confidence in each other)

Central to NEC3 is the principle that all parties agree to act 'in a spirit of mutual trust and co-operation' and this looks set to continue with the publication of NEC4 due out in June 2017. This requirement has been interpreted by some as introducing partnering into NEC3 and certainly there is a consistent emphasis on collaborative behaviour through NEC3's terms.

Partnering obligations can however be more specifically incorporated with the use of Secondary Option clauses (Option Clause X12) where partnering principles and repeat working arrangements can be demonstrated over time.

NEC3 focuses on early contractor involvement and front loading of work with a view to ensuring the best solution for the project and maximum value. This is demonstrated by the importance placed on the detailed development of the Works Information and the participation of the whole team in the preparation of the material contained in it.

NEC3 does not expressly provide for nomination of subcontractors. Contractually, all subcontractors are treated as domestic subcontractors, whether or not they are named or otherwise prescribed by the Works Information. Currently, this may create difficulties for the contractor if the terms of business of a named subcontractor cannot be brought into line with those in the principal contract.

In NEC4, however, the new design and build Secondary Options aligning the core conditions better with the Professional Services Contract (which is often used to subcontract out obligations) and the new integrated and full risk sharing collaborative agreement, Alliance Contract, may overcome much of this supply chain disconnect.

Until now, lawyers and other drafters of this contract have been forced to make substantial changes to address this omission. Alternatively, parties have simply relied upon the contractor's ability to request instructions from the project manager so that changes may be made to the Works Information as required.

Even so, novation is still not recognised within the suite because the process is considered incompatible with the NEC principles of good management practice. Similarly, the NEC3 Short Contract does not provide for novation or for nomination of subcontractors in its terms so we can expect certain amendments to continue even when using the NEC4 edition.

2.1.1.3 FIDIC

Although the Silver Book anticipates early contractor involvement in the design process, there is in many of the contracts a more traditional demarcation between design and construction.

Nomination of subcontractors is provided for at Part 5 of the Red Book. Part 5 sets out a procedure whereby the contractor may object to nomination and provisions regarding payments/evidence of payment to 'Nominated Subcontractors' (clauses 5.3 and 5.4). Responsibility for the nominated subcontractor is not expressly allocated. However, the Red Book does state that the contractor may object to nomination where the nominated subcontractor refuses to indemnify the contractor against any failure by the subcontractor to fulfil its obligations. The contract also states that: 'The Contractor shall be responsible for the acts of defaults of any SubContractor'. As a matter of interpretation, risk is therefore intended to remain with the contractor. The Yellow Book also provides for 'Nominated Subcontractors'. The provisions are, however, briefer than those in the Red Book.

The Red and Yellow Books do not provide for novation. Should novation be required bespoke amendments will need to be made to the standard forms. The FIDIC short form does not provide for novation or for nomination of subcontractors.

Interestingly, FIDIC proposes a new MBD (Multilateral Development Bank Harmonised Edition) version of the Red Book for Autumn 2017 (confusingly called, the Pink Book) which will introduce an additional ground of objection by the Contractor to a nominated Subcontractor where the Subcontractor does not accept that it will be paid only if and when the Contractor has received payment from the Employer for sums due under the subcontract.

In certain jurisdictions, including the UK, this type of pay-when-paid clause is unenforceable. In these circumstances, this ground of objection should fall away as the 'pay-when-paid' provision itself in the relevant subcontract will need to be amended to ensure it is enforceable in the jurisdiction in question and to ensure that quality is not compromised by a provision which is likely to cause much irritation and come across as a backward step to supply chains.

2.1.1.4 PPC2000 and TPC2005 (as subsequently amended in 2008)

The structure of the contract seeks to integrate the entire project team, who sign up to a single agreement. The contract is set up to cover the entire duration of the construction process from inception, feasibility and design to physical work on site. The overriding principle is that of collaboration, and the intended effect of the multi-party structure is that each party to the contract is contractually responsible to every other party. This also means that every party to the contract has the right to bring proceedings against every other party.

The integrated design, supply and construction processes include early design development with specialist contractor contributions prior to start on site (the constructor can perform services and other activities to this end under a Pre-possession Agreement). The objective is to allow the knowledge and expertise of the constructor and any specialist to be applied to the project at an early stage[1].

Within the team there is an emphasis on the joint selection of specialist subcontractors, suppliers and consultants ('Specialists'). The subcontracting provisions allow the client a significant degree of control over the selection process. Where the Specialist subcontractor is making a substantial contribution to the project he/she may be asked to sign the Joining Agreement and become a member of the partnering team.

There are provisions that allow the client to appoint a specialist directly. Responsibility for payment of the specialist remains with the client unless later appointed directly by the constructor. With the exception of specialists appointed under the terms of the contract, the constructor will be responsible for all aspects of performance of the specialist.

Given the multi-party structure and the fact that the key project participants all sign the same contract, amendments to allow for novation of

---

1 'Constructor' is a term particular to PPC2000. The equivalent in all other standard forms referred to in this reference guide is the 'Contractor'.

the design team should in principle not be required. PPC2000 does not recognise novation.

2.1.2 Appropriateness to the nature of the project

There is no getting away from the reality that certain contracts are more suited to particular activities than others. For instance, JCT was historically conceived with the built environment and construction in mind and this is where it sits best. Trying to use it to procure complex mechanical & electrical work to interface with specialist equipment for a process plant is possible but we end up fitting square pegs in round holes again. It can be done with thoughtful amendment but why not use something which has been conceived with that sort of project in mind, such as FIDIC Yellow Book, the MF series published by the IET or even the IChemE (chemical engineering contract) which has inherent process-driven type checks and balances already in place?

Dealing with just real estate alone, the types of activities of a particular project are far-ranging and this is one of the inherent risks that was discussed in Part 1.

For instance, the works could involve the construction of a completely new detached building or buildings (with or without phased completion), an extension to an existing building, a refurbishment of part or whole, the restoration of a historic structure, the reinstatement after fire damage or neglect, or may involve the demolition or a repair and maintenance programme involving many buildings. Each will have its own specific considerations in terms of cost, quality required and timescales for completion.

The question is how do you know which contract is most appropriate?

2.1.2.1 Suitability according to description

Standard form contracts tend to be tailored by institutions/bodies for a particular use and are often entitled as such, which makes it easier to make an initial choice.

For instance, minor works or major works contracts published by the Joint Contractors Tribunal (JCT) are supposed to be used for exactly that purpose. The contractual procedures are deliberately limited to allow for more flexibility and, arguably, a lighter and more proportionate touch. If either of these types of contracts is identified as being suitable for other reasons, then their terms may benefit from very specific revision for the requirements of a particular project.

Some contracts, for example, the Institution of Mechanical and Electrical Engineers and the Institution of Chemical Engineers deal more with supply chain type issues as is obvious by their titles. Whilst not dealt with here, lessons can nevertheless be learnt from those forms when contracting if more prescriptive procedures need to be put in place to align cost strictly with output, or if quality is such a critical factor to require certain testing and commissioning of installations prior to client handover.

The FIDIC forms of contract especially distinguish themselves in this way and are known colloquially by the colour of their covers, and are very specific to the nature of the project to be undertaken:

- Red Book (construction);
- Yellow Book (electrical and mechanical plant and DB engineering works);
- Silver Book (turnkey or engineering procurement and construction contract, where the extra cost is expressed as a premium for the contractor doing everything up to and including handover, e.g. power station);

- Gold Book (design, construct and operate, which represents an extension of services beyond completion of the works);

- Green Book (short-form of the above only to be used by hands on employers);

- the Blue Book (dredging and reclamation works); and

- a form known as the 'MBD', which is a pre-agreed form with certain global and pan-regional banks (the Pink Book harmonising the Red Book due out in Autumn 2017 being an example)

Other contracts, such as those published by the NEC, are more distinguishable by the way in which their pricing structure is set out, ranging from lump sum through to full cost reimbursable by a series of Options A–G, explained further below.

2.1.2.2 Suitability according to the intended purpose

Although generic in description and subject to exceptions to the rule, the following is a useful overview of potential uses (a full breakdown of the possible uses for each form of contract can be found in the table at the end of this section):

2.1.2.2.1 JCT

As mentioned above, the JCT has long been associated with the built environment. Its structure is understood across the industry and possible uses range from commercial property, both one-off and schemes with multiple completion dates, through to ancillary works such as road and services infrastructure for retail parks. Added since revision 2 in 2009, (we are now using the 2016 edition) JCT contracts are suitable for any project consistent with the principles of collaborative working, value engineering, sustainable development and increased environmental considerations. It can be aligned to measurement of success by the introduction of key performance indicators with increased project monitoring to ensure that those indicators are being achieved.

The downside is that JCT has been criticised in the past as entrenching parties due to the plethora of notice requirements. This view is somewhat unfair as the procedures contained within the standard form largely bring clarity to the parties' dealings with each other.

At times, the prescriptive nature of the notice requirements has however discouraged parties from acting commercially due to the mistaken belief that they may waive or prejudice their rights in some way at a later date if they do so. Alternatively, parties sometimes treat the contract form as a necessary evil to satisfy funding requirements and then do not refer to it again unless they need to at a later date. This has led to a certain amount of criticism that the terms of the standard form do not always reflect the activities on site, leading to certain amendments and the search for alternatives that match operative practice. This criticism, of course, can equally be applied to other standard form contracts.

2.1.2.2.2 NEC

The New Engineering Contract (NEC), now in its third edition and soon to be fourth, lends itself to most multi-party types of project, whether in the UK or overseas, due to the following factors:

- the strong collaboration ethos required to be exercised by all parties;
- responsibility by either party for design;
- a choice of pricing mechanisms from lump sum to cost plus;
- clarity and simplicity (which may also be its downfall); and
- good risk management processes including a largely effective early warning mechanism governed by a proactive project manager with decision-making responsibilities.

These factors can often make the NEC an attractive option for infrastructure and utility works. This is not its sole application however due to the flexibility afforded by a number of Secondary Options offering parties the ability to adapt and build up the contract in accordance with particular project specifics and employer and funder drivers.

Success in using the NEC suite requires a degree of pragmatism as it is built up using a number of primary and secondary options including:

- 9 core clauses;
- Currently 15 secondary option clauses;
- Bespoke terms or amendments from certain published or entirely ad hoc Z clauses; and
- A choice of two different schedules of cost components (depending on the main financial option chosen).

The NEC may not be suitable for all projects however, especially those where the high level of administration generated by operating the contractual terms would be disproportionate to the complexity or value of a project. An employer may not necessarily get best value from using NEC to build a single house, or even a small scheme of pre-designed houses, where the construction element is relatively straightforward.

Indeed, the mismatch (using any of the contractual forms) can and often does prove costly when parties either (i) fail to understand the importance of operating contractual terms in the correct manner, (ii) attempt to shortcut them in practice, or (iii) treat the contract in a different way than drafted and intended.

It is not uncommon to see established organisations attempting to run an NEC-based contract in the same way as it would a JCT one. Each requires a different mindset. Provided clients and their advisers are alive to the differences however and ensure that staff are adequately trained and knowledgeable about the processes in NEC3, this should minimise instances where the contract is inadvertently not agreed at all, or at least not in the way parties intended.

The potential problems in drafting that could arise include any of the following:

- A lack of attestation clause affecting insurance cover and the duration of parties' intended obligations;
- Vagaries in the drafting of the scope of services;
- The employer's risks being overlooked in the contract data. This can cause conflict as the contractor bears the risk of any not so stated;
- Indecision in respect of risk identified in the risk register and how to deal with them (in the standard form, there is only an obligation to record risks in the register; not necessarily to implement a course of action to reduce or remove that risk); and
- A failure to appreciate that non-compliance with certain notice periods can operate to bar entitlement to time and cost relief in respect of matters for which the contractor believes it is not responsible.

Much of these issues can be dealt with by robust internal systems to ensure that the contractual provisions are utilised to maximum effect. If these limitations can be addressed and minimised, the benefits of using NEC3 can be considerable leading to cost savings and success in terms of end delivery.

In addition, it may be appropriate in limited circumstances to use certain amendments, otherwise known as Z clauses, as these can help to clarify some of the obligations or project requirements. Although not dealt with in any great detail here, the following may need to be considered when using this form:

- The provision of collateral warranties, as discussed in Part 1, to comply with certain funding or other multi-party requirements;

- Copyright – where a contractor has design responsibility, it is important to include a provision addressing matters relating to copyright. Without such clarity the employer's rights, if any, to use the design may be rendered too uncertain;

- Prohibited materials and Codes of Practice – where a contractor has design responsibility, a provision requiring the contractor to refrain from using prohibited materials in design and construction may be necessary so as to prevent the use of unsuitable materials;

- Assignment – a provision prohibiting the contractor from assigning the contract (or limiting such assignment to a number of occasions) may be necessary so that the employer knows for certain who is actually carrying out the works;

- Confidentiality clauses – this may be of particular importance in the public sector or where a project is likely to attract certain media coverage;

- Project-specific definitions – to ensure that technical terminology is reflected in the document or for other environmental or cultural reasons; and

- Notification to the contractor of any third party interests, e.g. a landlord or funder, so that it does not act inconsistently or inadvertently put the employer in breach of its obligations. If the contractor cannot comply, then it is best to discuss this upfront before finalising contractual terms so that alternative provisions may be considered to get around the problem.

What should be remembered is that Z clauses are not intended to operate as a substantial re-write of the clauses of standard contracts. Indeed, the improper use of Z clauses can be problematic in itself.

2.1.2.2.3 FIDIC

The contracts published by the Fédération Internationale d'Ingénieurs-Conseils (FIDIC, otherwise known in English as the International Federation of Consulting Engineers) have been popular for all types of engineering projects for over 25 years, in particular those that are project financed.

Its use has, historically, been more international than being the contract of choice within the UK, but there has been a resurgence of interest in the last few years as UK businesses look increasingly to an overseas market for profitable work, most notably in North Africa and the Middle East. Moreover, it has been one of a number of forms of EPC contract (engineering, procurement and commissioning) which has been favourable to UK institutional funders for renewable energy projects in order to safeguard the quality of the underlying asset for the secondary refinancing market, which certain renewable energy projects now find themselves in.

The form is typically favoured by lenders and funders because performance of the asset to be built is typically critical to the lender's security and, therefore, project costs can be more easily ascertained. The structured, pre-defined tender process within FIDIC leading to the contract award and agreement should give parties and their advisers adequate time to undertake any due diligence that may be required.

Funding is normally linked to the ability to operate and generate revenue from the resultant building or facility, e.g. by a power station selling electricity. Therefore, there is much emphasis in FIDIC on testing, commissioning and handover procedures and the use of key performance indicators to measure time, cost, designed production and output levels. Full completion is not deemed to have been achieved until final performance tests have been carried out and the results proved positive.

FIDIC tends to place all design responsibility upon the contractor (or constructor) insofar as it is tasked with ensuring that the end result is 'fit for purpose'. In the context of the turnkey version (the Silver Book), the constructor is tasked with achieving the performance of whatever is necessary for a certain purpose (e.g. the particular yield of power generation) and so enjoys a greater degree of flexibility but also a higher risk if things go wrong.

It is not however the case that all risk is transferred to the turnkey constructor in FIDIC. Many large projects can involve several specialist turnkey constructors and the cross obligations with the employer can be complex. In other versions of FIDIC, there are several carve outs to this rule also. For instance, in the Yellow Book the constructor is not responsible for errors/omissions found in the employer's requirements and can normally expect this to result in a variation if amendment is required.

FIDIC may not however be suitable in the following circumstances:

- Where there is insufficient time for checking relevant information and scrutinising the employer's requirements. This is not a contract to 'bounce' on constructors at the last minute in the hope that competitive, 'best price' tender bids are then received;

- If the works involve a lot of underground work or areas where the constructor cannot verify the physical conditions;

- If the employer needs to supervise closely or inspect the drawings throughout the build and installation;

- If interim payments require an official or other intermediary to certify them; or

- Due to cost implications, the suite may not always be commercially viable for smaller projects.

Some care may also be required when tailoring the contract in terms of definitions employed, especially if these are to be translated into other languages at a later date for cross jurisdictional projects.

Although the form uses everyday terminology, some of the references can be misleading as their definition can be more or less defined than normal usage. For instance, the use of 'plant' is contractually defined as being wide enough to encompass not only plant in the normal sense but also apparatus, machinery and even vehicles that will be incorporated into the permanent works. Given the amount of defined terms in FIDIC, interpretation of definitions can be a trap for the unwary and so all terms should be double-checked as part of any contract/tender review process.

Definitions can also be worded so that they are dependent and effective upon a certain amount of knowledge being in a party's possession at a particular moment in time or conditional upon an independent act. This sometimes requires further amendment for the sake of certainty.

Other matters, not dealt in detail here but which often require additional wording for clarity in the FIDIC forms, include:

- The timing of various communications including a mechanism for evidencing delivery as well as a list of those that need to be notified for the communication to be effective contractually;

- Clarity as to which law/laws and language/languages are the ruling laws and languages for multi-jurisdictional projects;

- Certainty as to when access to the site will be given by the employer and if any restrictions or staged access for parts of the site need to be put in place having regard to the various places of work that may qualify as forming part of the 'site';

- limits as to who comprises 'employer's personnel' so that the contractor does not inadvertently benefit from certain indemnities that were not contemplated by the parties at the time of contracting; and

- where multiple contractors are on site, ensuring that their respective rights to claim from each other in respect of costs resulting from delay and failure to co-operate are preserved.

FIDIC works where there is a strict performance requirement which demands testing and commissioning. It also sets out useful project management stages for running tenders and, for this reason, its transparency is attractive to the public sector which tends to adapt it to local laws and usage.

2.1.2.3 Suitability by proximity, knowledge working relationship of the parties

Over the last 15 years or so, there has been a push for parties to work collaboratively or to 'partner' with each other. These concepts are sometimes misunderstood, as a gathering body of case law in England & Wales has demonstrated. There is no doubt that working in collaboration to generate mutual trust and confidence is a positive aspiration and many projects have benefited from doing so to achieve spectacular results.

Partnering does not however mean that contractual structures or terms are watered down or ignored. In fact, in such cases the need for contractual clarity is most important to avoid misunderstandings as to who performs which role and when. Partnering principles are enshrined in a number of forms of contract including:

- JCT Constructing Excellence;
- JCT 2009 editions onwards (including the current 2016 edition) incorporating collaborative working;
- NEC3 generally but enhanced by Option X12 Partnering; and
- ACA PPC2000 and TPC2005 (as amended in 2008).

Certain factors may mean that partnering is not appropriate or that a relationship needs maturing before that level of commitment can be given by parties from an organisational point of view. It may be that a partnering arrangement, which requires a significant amount of interaction between parties or core groups, does not represent value for money in terms of the professional fees expended for a more modest project.

Similarly, where a one-off, more straightforward project with, perhaps, an overseas investor at arm's length is involved where a tight programme is envisaged, then partnering may also not be the most effective option.

Even for repeat projects, it may be that the parties can streamline the process further whilst still agreeing core principles surrounding expected levels of quality, time frames and cost by entering into a framework agreement, (such as those published by JCT and NEC) which then acts as an umbrella agreement with individual, short-form contracts then 'called off' at pre-agreed key dates or as and when required. Alternatively, the parties may like to consider a measured term contract as a precursor to entering into either of the above to test the water over a period of time.

If the circumstances of the parties mean that partnering may be an aspirational or already a very real and effective way of proceeding, then the following is a brief insight into some of the global considerations that need to be borne in mind if using any of the partnering based contracts above.

2.1.2.3.1 Considerations in partnering between parties to assure the core objectives

It is often the case that the contract lists those parties which have agreed to work to a policy of co-operation with, perhaps, on larger projects a core group normally representing the wishes of all members.

It is often wise to list the types of activities expected by those participating in the partnering information against a master programme to promote accountability of each of the partners. This tends to work especially well when coupled and measured in accordance with the achievement of certain targets and continuous improvement, for the reasons set out below.

Often, the concept of 'partnering' can be opaque in its meaning and disputes can occur leading to overruns in terms of time and legal costs. The simple problem is determining precisely what the term means.

Some authority is given for the proposition that 'to partner' is akin to a requirement to acting in good faith, which is a useful comparator when operated in jurisdictions which have codified good faith provisions as part of their legal system. This does not work so well in England & Wales, which has a common law system, although a number of cases have tried to make sense of the concept of partnering in that jurisdiction[2].

In England & Wales, partnering does not necessarily imply that parties must avoid acting in such a way which *would* 'damage the relationship or confidence in trust' between them. If the terms of the contract are quite specific as to what level of interaction is sufficient to discharge the duty of care to one another 'to partner', then the courts may be more inclined to offer opinion upon any breach of such terms.

In spite of judicial guidance, partnering seems to work effectively and practically in a contract when it can be measured against a number of activities which suggest that a standard of care exists if a party acts in a certain way.

One thing is clear, being too prescriptive in terms can result in the parties losing the whole ethos of partnering, whereas not being prescriptive enough can lead to uncertainty and unenforceability of terms.

In addressing this problem, the English Court of Appeal has been ready to acknowledge a binding standard of care of partnering between commercial entities that have sought professional legal advice in setting out their relative obligations to one another, thereby evidencing the intentions of the parties and where the drafting is such that the cost of not acting in accordance with certain good faith provisions can be calculated by the court[3].

As a matter of drafting partnering provisions, it is often useful to agree how prescribed activities are to be measured as having satisfied the

---

2 see, e.g. *Bedfordshire County Council v Fitzpatrick Contractors Limited* (1998)

3 see, e.g. *Birse Construction v St David Limited* (1999); *Petromec Inc and Others v Petrobras and Others* (2006) CA

behavioral norms expected between the parties. This is not without difficulty as it usually requires agreement as to the method of measurement between the estimated and actual target to be achieved, clarity in respect of the specific target to be achieved and a mechanism for any bonus and how it is to be paid in some sort of bespoke 'incentive schedule'.

Alternatively, if the parties are not ready to commit to a binding partnering arrangement, then the spirit of collaborative working could be reflected in a separate pain/gain mechanism arrangement if certain targets are achieved or not, such as can be found in Option X20 of the NEC suite.

If you want a fully-fledged partnering contract, then the PPC2000 and TPC 2005 (as amended) forms are exactly intended for that purpose albeit it is still true to say that partnering is, arguably, a matter of conduct; not contract.

We have all seen partnering charters, good faith provisions and other expressions of mutual appreciation. The reality is that a lot of it is aspirational and not contractually binding. In the UK, the case of *Birse Construction v Exeter St David Ltd [2000] ABC.L.R. 08/17* at para 37 onwards will give you a flavour of how badly parties can get it wrong when it comes to incorporating partnering arrangements into contracts.

So the question is: do I use NEC or PPC2000/TPC 2005 if I am looking for a highly collaborative team? The answer really depends on the team itself as well as certain other practicalities.

When it comes to UK local authorities, they are great advocates of NEC3 with the partnering option weaved in. This is not a problem provided the exact activities indicating what constitutes "partnering behaviour" are clearly defined and risk is apportioned appropriately between those parties who are best placed to bear it.

In facilities management particularly, NEC3 comes highly endorsed and for good reason[4]. Provided everyone understands how it works, and

4 Construction Clients' Board and Facilities Management Board of the Cabinet Office in Achieving Excellence in Construction.

the project management team are a pro-active bunch, it is an effective procurement route for multi-party, complex projects. Problems only tend to arise with NEC3 when the parties significantly amend the standard form to the point where the ethos of contracting NEC principles (and the partnering provisions) start to get eroded.

There are a number of benefits to TPC2005 however, which are worth listing for the sake of completeness and balance. These include it:

- being a natural fit for Building Innovation Modelling (BIM) meaning better information and potentially lower administration for ongoing maintenance if TPC is used (although it will be interesting to read the new BIM provisions in NEC4 which purport to do something similar);

- allowing more readily, without significant amendment, the provision of 'call off' contracts with Specialists on a rates basis. NEC3, by contrast, is more appropriate where there is a precise amount of work identified at the outset defined in the Service Information for which the Contractor is paid, as opposed to work instructed by the Employer, as and when required;

- avoiding much of the 'Z' clause amendments required to adapt the NEC to project specifics. The paradox of NEC has always been that it is written in plain English but, at times, to the detriment of contractual certainty without substantive amendment to the wording which can itself lead to even more uncertainty unless drafted very carefully and with all the consequences being considered upfront;

- minimising discrepancies/errors from arising between contract documents due to the establishment of a Core Group of relevant parties, whose role it is to review and 'stimulate' the programme, including dealing with any proposals for remedying any matter which threatens the progress of the services. In this way, TPC2005 goes a step further than NEC in the even closer administration and pro-active management of the quality of the services provided;

- Allowing contracting parties more readily, via the open book accounting embedded in the TPC2005 way of working together, to demonstrate substantive cost savings and streamlining of systems overall; and

- Encouraging triple bottom line accountability of the Contractor in its obligation to employ sustainable and continuous improvements in its processes as well as value management and value engineering practices.

The current balance of favour is probably with PPC2000/TPC2005 due to its intuitive nature but, as mentioned above, NEC is hot on its heels with its new edition which can only be a good thing for the concept of evolved partnering.

2.1.3 Control by appointed third parties

2.1.3.1 The JCT approach

The majority of JCT contracts benefit from the requirement to have a robust contract administrator in place to ensure that the works are carried out and completed in accordance with particular project requirements. As the table below demonstrates, certain contract forms also benefit from having a quantity surveyor (or estimator) on board as well where the contract is set up so that a certain way of pricing is paramount to assure cost is kept in check.

The existence of these consultants is not however fool-proof; the success of a project can still be prejudiced by over-certification or signing off elements of the works as practically complete when they should not be, often in the mistaken belief that this will help keep the parties working together on site. The outcome can be quite different than envisaged.

Where not stated, the various contracts under the JCT do not have either architect/CA or quantity surveyor requirements:

Contract	Architect/CA	Quantity Surveyor
Standard Building Contract	Yes – all versions	With Quantities version With Approximate Quantities version
Intermediate	Yes	No
Design & Build	Called 'Employer's Agent' and wider powers	No
Major form	Called 'representative' and exercises Employer's powers	No
Construction Management	No – the construction manager manages the trade contracts on behalf of the employer	No
Management Building Contract	No – the management contractor administers the conditions of the various works contracts	No
Measured Term	Yes	No
Prime Cost	Yes	Yes
Home Owner	With and without	No

Typically in a JCT-based contract, the architect/contract administrator will have a dual role of acting as the agent for the employer and acting as an impartial certifier and is subject to a number of very prescriptive functions aimed at ensuring the procurement proceeds smoothly or there is a detailed mechanism for when change occurs or things go wrong. To that end, it is typical to see an architect/contract administrator involved with ensuring that:

- The contractor complies with its instructions in relation to the integration of any design;
- Materials, good or standards of workmanship are of sufficient quality;
- A copy of the contract documents is kept on site so that the architect/contract administrator can refer to them in case of any issue arising;
- It gives the contractor accurate determinations of levels and correct dimensioned drawings with any other information required for the contractor to set out the works;
- It gives the contractor particular information in the form and at the time agreed in advance;
- It issues further drawings and details to amplify contract drawings and issues instructions as necessary to enable the contractor to carry out and complete the works including instructions to overcome any discrepancies and variations to the scope of contractual works;
- It ensures that no materials are moved off site without good reason;
- It manages any delay with the contractor and assesses whether the contractor qualifies for time relief in certain prescribed circumstances stated in the contract;

- It has wide powers of access to the works and the contractor's other sites where work under the contract is being done;

- It can govern the basis upon which the contractor subcontracts out any works;

- When practical completion of the works has been achieved, it can recommend the release of half of the retention and ensure cash flow to the contractor;

- Any increase in the contract sum is undertaken in accordance with detailed procedures for agreeing such increases whilst ensuring that the contractor's cash flow is maintained by a system of interim certification;

- It issues a notice of non-completion if the contractor is late in completing, which is a condition to the employer being compensated for any costs incurred (such as lost rent) in the overrun. This power may also incentivise the contractor to complete on time;

- It can exercise its power to allow the employer early or partial possession to alleviate budgetary and time constraints upon the employer, whilst not penalising the contractor for being unable to assure the requisite quality of the project as envisaged;

- It notifies the contractor of any defective works and any other failure to comply with design obligations; and

- It certifies the making good of any defects, thereby releasing the remainder of the retention to the contractor as well as ensuring the quality of the end product.

2.1.3.2 The NEC approach

Similarly, the NEC suite is governed heavily by the role of the appointed project manager who undertakes a proactive approach to administering the contract terms as follows; namely, by:

- Playing a collaborative role in conjunction with the employer and contractor;
- Accepting the contractor's design and programme only when happy for it to be part of the contract;
- Communicating with the contractor in prescribed ways and within prescribed time frames;
- Providing instructions that may vary the scope of the works or working areas and may vary time frames for completion in certain limited, prescribed ways;
- Actively monitoring the works by means of an early warning mechanism in case of anything that increases the price, delays completion, delays meeting a key date or otherwise impairs performance of the works in use;
- Instructing the contractor as to how it should overcome any event of prevention in carrying out and completing the works including the issue of instructions to accelerate;
- Assuring the quality of the subcontractors to be used;
- Certifying the works as being complete;
- Allowing for adjustment of the prices to allow for any change in the works information or any uncorrected defects;
- Assessing and certifying amounts due to the contractor at prescribed times;

- Notifying the contractor of the outcome for any claim for a compensation event and requesting quotations for any proposed instruction or changed decision;

- Issuing any defects certificate; and

- Agreeing any extension for referring any dispute to adjudication.

The NEC also benefits from a 'Supervisor', which undertakes a monitoring role and thereby assures quality at the end construction by reviewing any defects. This power extends to instructing the contractor to search for a particular defect and stating, in the defects certificate, what defects (if any) exist[5].

2.1.3.3 The FIDIC approach

The hallmark of FIDIC is balanced risk and it is monitored, in most forms, by an experienced person known as the engineer, who does the following:

- Issues instructions to deal with instances of delay and disruption, remedying of defects and instructing any variation;

- Obtains employer authority for matters pre-agreed between the employer and contractor as necessary. The engineer benefits from ostensible authority to act on behalf of the employer in most instances;

- Exercises an extensive and central role to the project and has the power to decide matters, extra-judicially, provided such 'determinations' are made on a fair and reasonable basis;

- Maintains control over any replacement of the employer's representative;

---

5 Further guidance can be found in the NEC publication *Choosing the right NEC contract*' published by Thomas Telford Publishing), first published in 2006

- Maintains quality control as the person to whom the contractor sends all design information, procedures and compliance documents and attends inspections and testing of works prior to covering up, as and when required;

- Is responsible for the taking over process for completion of the works or sections of it;

- Certifies applications for payment but only once any requirement to have financial security in place has been discharged by the contractor;

- Plays an active part in the dispute resolution process, can assure the protection of written records to aid contemporaneous evidence for its own determination of the contractor's claims and/or provide fuller information by consideration by a Dispute Adjudication Board.

2.1.3.4 Considerations about appointed third parties when assuring the core objectives

Quite often, appointed third parties under a contract have dual roles when employed by the employer – an agency role and contract administrative role. Sometimes these roles can be seen to conflict.

Although by no means limited to FIDIC, there is a familiar paradox concerning the role of the engineer as it is employed by the employer but is obliged to make 'a fair determination in accordance with the contract, taking due regard of all relevant circumstances'.

This tension between the dual roles has generated a certain amount of case law concerning certifiers 'fettering' (or compromising) their impartiality. The current position has culminated in the acceptance that the duty to act impartially only arises when the appointed third party is administering the terms of the contract or acting in its 'decision-making function' as opposed to its 'agency function', when its role is to advise the employer in its best interests[6].

6 Scheldebouw BV v St. James Homes (Grosvenor Dock) Ltd [2006] Adj.L.R. 01/16

What is fair does, of course, depend on the particular circumstances. The fact that the appointed third party may only give the contractor one day to consider how to overcome certain delays or respond to complex technical allegations, may not in itself be unfair if there is another pressing time reason which threatens the successful completion of the project and justifies this instruction (*AMEC Civil Engineering Limited v Secretary of State for Transport* (2005 CA).

One thing is clear however, the engineer under FIDIC contracts may enjoy more independence in decision-making powers than its counterparts under the JCT or NEC suites as it is not required to apply a judicial-type approach in its decision-making process. Nevertheless, the engineer is still required to consult with the employer whenever it needs to make 'a fair determination in accordance with the Contract'.

It may be possible for the parties to relieve this inherent tension by agreeing a list of decisions for which approval is not required or pre-given. That aside, if the reasonableness of an appointed third party is called into question too many times, it is likely that the parties will need to address it by other means whether by means of negotiation, mediation, adjudication, litigation or, specific to the FIDIC forms, by engaging the Dispute Arbitration/Review Board tasked with dealing with issues on site as they arise[7].

There are subtle differences as to how administrators of JCT, NEC3 and FIDIC interpret and operate their duties. The following provides a brief oversight and non-exhaustive list of some of the differences in that may arise when moving between the standard forms.

A) Conclusivity of decision making

In JCT, the Contract Administrator is afforded a certain amount of protection in terms of its decision making responsibilities. Provided the decision is 'reasonable' then it should not be open to successful challenge. The standard provisions make the final certificate conclusive in the following respects; namely that:

7 Dispute resolution mechanisms contained in the contracts are not covered here in any detail, but numerous titles are available elsewhere considering the options available.

- The quality and standards are to its reasonable satisfaction;
- Additions, omissions and adjustments to the Contract Sum have been correctly made save in the instance of accidental inclusion or exclusion or arithmetical error;
- All and only such EOTs contained in the certificate have been given; and
- All reimbursement of direct loss and expense has been made to the contractor.

This is, of course, subject to a party not bringing an action within 28 days of the certificate being issued, in which case it is only conclusive in respect of those matters not subject to that action. No other certificate of the Contract Administrator is conclusive that the Contractor's works, design, materials or goods are in accordance with the contract.

In comparison, NEC3 does not have express conclusivity provisions. It does however contain a mechanism in terms of assessing additional time and cost, whereby the Project Manager is deemed to have accepted the Contractor's proposal if it has not rejected it within a specified time-frame. For instance, in condition 61.4, if the Project Manager does not notify his decision to the Contractor as to whether or not an event constitutes a Compensation Event either within one week of such a request (or such other period as the parties decide) then a failure by the Project Manager to reply within two weeks of this notification amounts to acceptance that the event is to be treated as a Compensation Event. This is also deemed to be an invitation for the Contractor to submit quotations. Although not a conclusivity provision as such, it does mean that any adjudicator/court should find in the Contractor's favour should a debate as to the status of an event subsequently arise.

Similarly in FIDIC, there are no express conclusivity provisions, however, the parties are required to give effect to each agreement or determination by the Engineer pursuant to clause 3.5 "until and unless revised under clause 20 [Claims, Disputes and Arbitration]."

B) Subjective decision making

This is an inherent issue in all of the contracts, where the subjective opinion of the Contract Administrator could be called into question, especially as in JCT and NEC3 where the relief for the contractor may hinge on the certifier "becoming aware" of a certain event or circumstance having taken place.

In JCT forms, the contract stipulates that "if" the Contract Administrator becomes aware of any divergence between statutory requirements and the contract documents, it shall give an immediate notice to the Contractor. Within 7 days of becoming aware of such divergence, the Contract Administrator issues instructions which (if they vary the Works) amount to Variations.

In JCT, this subjectiveness in becoming aware is limited and not extended to wider matters. Unlike NEC3 below, it is only 'if' the Contract Administrator becomes aware that the clause is triggered, so a Contract Administrator may never become aware of any divergence until and unless a situation arises. To overcome this, the Contractor is not prejudiced and can obtain relief in clause 2.17.3 so that it will not be penalised for carrying out non-compliant works that are stipulated in the contract documents.

In comparison, the NEC3 form says that the Project Manager shall notify the Contractor of any matter which could increase, delay or impair the Works "as soon as [it] becomes aware". This is worded much wider than in JCT. However, any ambiguity in the wording of the contract is intended to be tempered by some inherent fail-safes. If, for example, the Project Manager fails to notify the Contractor of a Compensation Event when it should have done, the Contractor will not necessarily be time barred from applying for relief for time and additional cost (condition 61.3).

There should be less scope for the same risk to arise in FIDIC which requires the Engineer to obtain the Employer's consent when exercising a specified authority. The requirements and situations for obtaining consent should be pre-agreed and listed in the Particular Conditions (clause 3.1). In all other respects, the Engineer should be free to make determinations on a fair basis (clause 3.5). Therefore, there is arguably more certainty as to when the Engineer can and cannot act of its own accord.

C) Delegation of powers

JCT forms state that any person authorised by the Contract Administrator should have certain rights such as, for example, access to the Works. Whilst in all contracts replacement of the Contract Administrator is discouraged to avoid uncertainty and disjointedness in the procurement of works, there is a 21 day period for the Employer to nominate a replacement and to give the Contractor notice. To maintain consistency, that replacement is then bound by any certificate, decision instruction etc made by its predecessor, unless they have the contractual power to disregard it.

The ethos of FIDIC is that it is a neutral form of Contract. It is possible for the Engineer to assign duties and delegate authority (cl 3.2) but may only delegate his duty to make 'fair' determinations with consent of the parties.

In NEC3, however, the Project Manager may delegate authority with the only requirement being that it first notifies the Contractor. There is no Employer control mechanism here.

D) Meetings

Site meetings are pretty much at the Contract Administrator's discretion in JCT. In NEC3, the procedures for early warning and problem solving are much more prescriptive and risks need to be recorded. In FIDIC, the Engineer has the discretion to attend management meetings as it sees fit.

E) Information flow

Unless the parties agree in JCT that an Information Release Schedule is in place, the obligation on the Contract Administrator is simply to provide copies of contract documentation upon execution of the contract (clause 2.8).

In NEC3, the Project Manager may be part of the team that puts together the Works Information. Thereafter, the onus is on the Contractor to provide the Project Manager with documentation to show how it proposes to 'Provide the Works' and any revisions thereof submitted for acceptance.

In FIDIC, the parties will have obtained much of the information required throughout the prescriptive tender process. Thereafter the Engineer may issue to the Contractor (at any time) instructions and additional or modified drawings for the execution of the Works (clause 3.3). It may be that the Engineer also asks the Contractor for proposals prior to instructing or rejecting a Variation (clause 13.3). The information flow is therefore more fluid than perhaps under the other two forms.

F) Practical Completion

The concept of practical completion does not exist in FIDIC. Instead, there is a procedure for 'Taking Over' the Works by the Employer following a period of commissioning. It is up to the Contractor to apply to the Engineer for a Taking Over Certificate.

Thereafter, the Engineer has 28 days to reject or issue the certificate. Importantly, if the Engineer does neither, then the certificate is deemed to have been issued on the last day of that 28 day period.

In NEC3, Completion is when the Contractor has done all the work which the Works Information says he is to do by the Completion Date and when it has corrected notified defects which would have prevented the Employer from using the Works.

The process of achieving Completion itself is subject to a period of testing led by the Supervisor, employed by the Employer, who is the ultimate arbiter (not the Project Manager) of whether or not the Works are defect free. Importantly however is the fact that the Contractor remains liable for correcting defects whether or not the Supervisor notifies it or not.

In JCT, however, it is (as a general rule) the Contract Administrator's opinion that counts as to when practical completion has been achieved provided certain design documents (if required) and CDM requirements have been complied with.

This can, on certain projects, result in the parties agreeing and inserting a definition of practical completion in order to avoid argument at a later date as to when it occurred or should have occurred. Ultimately, practical completion is a matter of fact and the contractual definition serves only to trigger release of half of the retention being withheld from payment, if retention is applicable to a particular project.

As with any contract, standard industry or bespoke, the parties need to consider the precise role of the parties. Contract administration can substantially aid the successful completion of a project provided everyone understands the precise scope of responsibility and power afforded to the person undertaking that role.

2.1.3.4.1 Summary of scope and main responsibilities under each form

JCT Contract Administrator	NEC Project Manager	FIDIC Engineer
Overall		
- Administers the contract and attempts to assure Contract Sum by ensuring that the Contractor complies with its instructions. - Wider powers when acting as Employer's Agent.	- Proactive administration requiring skilled judgment. - Collaborative role with Employer and Contractor including prescribed communication routes. - Acts in a spirit of mutual trust and co-operation with the Employer, Contractor	- Deemed to act for the Employer and required to obtain consents as outlined in the Particular Conditions (cl 3.1). Obtains Employer authority for matters pre-agreed between the Employer and Contractor as necessary. Benefits from ostensible authority to act on behalf of the Employer

- In certain instances, contractual provisions can be subject to "fair and reasonable" assessment.	and the Supervisor (condition 10.1)	in most other instances. - Exercises an extensive and central role to the project and has the power to decide matters, extra-judicially, provided such 'determinations' are made on a fair and reasonable basis after attempting to consult with the parties (clause 3.5).
Quality control - Issues timely information relating to set out and any other pre-agreed information as and when required.	- Acts as gatekeeper to Contractor's design and programme forming part of Contract and use of subcontractors. considering and accepting the Contractor's design, if any (conditions 21-23), any subcontractors (condition 26) and the proposed programme (condition 31). - replies within the period for reply to any communication submitted to it by the Contractor.	- Issues instructions and additional or modified drawings as necessary for the execution of the works and the remedying of any defects (cl 3.3). - may assign duties and delegate authority (cl 3.2) but may only delegate his duty to make 'fair' determinations with consent of the parties. - May attend management meetings with the Contractor.

	- issues certificates to the Employer and Contractor. - operates the early warning mechanism by notifying the Contractor of any increase, delay or impairment to performance, entering such instances in the Risk Register, attending risk reduction meetings and making decisions and finding solutions with others to overcome those risks (condition 16). - issues instructions dealing with acts of prevention (condition 19).	- Contractor sends all design information, procedures and compliance documents and attends inspections and testing of works prior to covering up, as and when required.
Variations/change - Issues instructions/variations to overcome discrepancies and/or changes in scope of Works.	 - Issues instructions relating to changes in scope and completion date.	 - Issues instructions to deal with instances of delay and disruption, remedying of defects and instructing any Variation. - Requests measurement of the Works, if required (12.1);

- Issues notices in respect of remedying discrepancies between contract documents (cl. 2.13 to 2.18).	- Actively monitors by means of an early warning mechanism for any change to scope, price, timings or impairment of performance.	- Issues Varations (13.1) and payment certificates (14.6, 14.11 and 14.13); - Determines the contractor's claims (20.1).
Time and money - Certifies sums due. - Assesses delay and grants time and cost relief according to prescribed provisions but subject to overriding power to grant time relief as is fair and reasonable immediately following PC. - Issue extensions of time which it considers fair and reasonable and reach a decision as soon as reasonably practicable (cl 2.28.2). - Consider with the QS, if employed, all interim valuations, claims for variations and loss/expense	- Certifies sums due. - Decides the date of Completion and certifying Completion (condition 30). - Notifies the Contractor of the outcome for any claim for a Compensation Events and requests quotations for any proposed instruction or changed decision. - Assesses the additional cost of the Contractor not achieving a key date (condition 25.3). - Considers compensations events, their value and instructing their implementation (conditions 60-65).	- Certifies sums due. - Agrees the commencement date, the programme for the works (8.1 and 8.3), extensions of time (8.4), variations (13.1) any submissions for value engineering (13.2) and the sufficiency of the Contractor's payment security.

resulting and issue payment certificates as appropriate.		
PC and snagging - Determines when PC achieved and issues notice of non-completion, defects and can allow early possession. - Issue Practical Completion Certificate or Sectional Completion Certificates (cl 2.30 to 2.32) or a Non-Completion Certificate (clause 2.31); - After practical completion, immediately consider the overall position on extensions of time and exercise discretion as to whether any further adjustment to time should be made. - Address any actual or suspected defects and exercise its power to specify	- Determines when PC achieved. - assessing defects and their value (conditions 40-45).	- Responsible for the taking over process for completion of the works or sections of it. - Assesses the time for and issues a Taking Over Certificate upon application from the Contractor (10.1).

and require remedial action arising during the Rectification Period, preparing both a Schedule of Defects (cl2.38.1) and by issuing instructions for individual defects (cl 2.28.2) and/or making an adjustment to the Contract Sum if the Employer agrees that any defect may remain.		

2.1.3.5 Pricing structures

As explained in Part 1, there are various pricing structures which may be appropriate in any given set of circumstances. How a project is priced reflects the relative risk for one part or the other. The table at the end of this section highlights some of the factors that may need to be considered in more detail.

2.1.4 Choosing a contract specific to project requirements

One of the most important things to recognise is that there is no hard and fast rule as to which contract to use once one gets past the initial structure largely dictated by funding issues, timescales, scale/complexity and quality control. Indeed, the chosen published standard form then becomes a key document but other factors specific to the project may require that it be amended for a number of reasons.

For example, there may be a requirement for a party to provide key documents, obtain certain signatures or put in place certain funding arrangements, such as an escrow account, prior to entering into the contract, gaining access to the site or even getting paid.

The decision thereafter can often be as much to do with parties knowing and preferring particular contracts more than others and feeling comfortable operating those terms and being in control, than being based on the application of any objective criteria. The best contractual structures are ones that provide certainty but at the same time a degree of flexibility too so that they can be changed by agreement of the parties relatively simply and without excessive cost to changing circumstances.

There is however an inherent risk to note here. It is also true to say that the nature or set up of a project can radically change from that which was originally intended. A contractor may originally have agreed to undertake only a portion of design responsibility which then expands to the point where it has assumed, by conduct, a much greater design role than first envisaged.

In such circumstances, it is wise to review the contractual structure sooner rather than later and ask two key questions:

- Does this contract adequately reflect what is happening on site?
- Can we adapt the current contract to meet the changing requirements or should we reconsider our obligations and responsibilities and agree a new contractual structure?

The parties may prefer to stick with their existing structure, even upon advice to the contrary, often in the misguided view that it is more cost effective not to do so than bringing a contract to an end by mutual consent and drafting a new one. Consequently, a contract may be heavily amended to fit the changing requirements of the project when it may sometimes be better in the long term to bring the existing one to a close and enter into a more appropriate form, for instance, where a one-off project is then replicated across many sites.

Most forms of standard form contract have abbreviated versions. It would, for instance, be overkill to use a full version containing more detailed procedural aspects for a relatively straightforward development of less than, say, £300,000 unless there were complex works or a high degree of building services to be carried out.

As an example, the JCT has published Immediate and Minor Works Contracts used for more marginal valued projects. Typically, an Intermediate Form may be suitable for projects valued between, say, £300,000 and £1m. This may or may not include design obligations which can be less detailed than a full Design and Build Contract.

It is not uncommon for the parties to enter into a period of maintenance once the works have been completed. Such is often the case with properties which are then tenanted with a service charge covering such maintenance for mechanical and/or electrical installations. In such circumstances, it may be that a separate measured term contract provides a more helpful initial structure due to certain prescribed keystone dates which can then be defined.

Likewise, if multiple parties are to co-ordinate their services and have mutual obligations in respect of one another, then a project agreement with a framework agreement outlining the types of tasks and activities and the parties' responsibilities may provide better structure.

It may be that there are other factors that make a particular form more attractive than another, some of which are listed below:

- increased employer involvement in the construction process;

- the need to ensure that the construction itself is to a high quality finish, where the employer may wish to retain full design with its professional team;
- the employer's aspirations for an early start on site;
- a short or 'tight' programme; and
- phased, or sectional, construction as a key requirement, for instance, in commercial retail projects where a landlord may be keen to get its various tenants into units and trading as soon as possible.

The issue of design responsibility is used below as a practical worked example as to how the decision process may unfold, considering the relative merits of each contract against our aspirations of achieving a mixture of cost certainty, meeting deadlines and assuring quality.

2.1.5 Worked example – design issues

Questions:

- ***How much does design quality drive the scheme?***
- ***Does the client want flexibility of design change?***
- ***Is the design already quite developed?***

Several questions arise here which will influence which contract can be used and which is most effective.

If quality of design is a key driver, the employer may wish to maintain full design control – it being fully designed by appointed designers and a more traditional route followed. It may still be possible however and indeed appropriate to adopt a more hybrid approach so that some of the design is placed in the contractor's hands and its specialist subcontractors, in which case a design portion type model may be more appropriate. The employer may still wish to maintain control over the selection of specialist subcontractors, the choice of materials and workmanship in which case a management form of contract could be considered. It may be that certain performance specified requirements can measure quality and that this is an important feature.

The employer may however be looking for a more integrated approach with single point responsibility, meaning that a design and build or turnkey approach may be favoured. Obviously, each option will have consequences on price and the flexibility to change such design decisions in the future. With that in mind, what are the options available?

The range of possible options is as follows:

Type of contract	Who designs?	Ease of use
JCT		
JCT SBC with or without quantities	Employer, but can include contractor design portion	***
JCT DB	Contractor, but employer still has to state its requirements	**
JCT Intermediate Contract	Employer, but can include contractor design portion	***
JCT Homeowner with or without consultant	Employer	***

JCT Repair and Maintenance	Employer	***
JCT Pre-construction services	Design team including contractor	***
NEC		
Engineering and Construction Contract	Employer, contractor or hybrid	**
Engineering and Construction Short Contract	Employer	*
Professional Services Contract	Consultant	**
Engineering and Construction Sub-Contract	Subcontractor for the subcontracted element only	**
Engineering and Construction Short Sub-Contract	Employer or contractor	*
Term Service Contract	Employer	*
Framework	Employer or supplier	***
FIDIC		
Red Book	Employer with option of a contractor portion (cl. 4.1)	***

Green and Blue Books	Employer	*
Silver, Yellow and Gold Books	Contractor	**
ACA		
PPC 2000	Design Team (which may or may not include the contractor)	**
SPC 2000	Specialist subcontractor with design team integration at early stage	**

*** = straightforward

** = some consideration needed

* = detailed consideration needed

2.1.5.1 Design in JCT

As JCT recognises itself, it is during the pre-construction period that the employer is able to derive most benefit from value engineering exercises and the input of others in the design process including, arguably, the contractor. The pre-contract services agreement can be used to most effect here.

Nevertheless, and more often than not, the design may evolve solely from the design team engaged by the employer even in outline as a result of concerns about engaging with a particular contractor and consequently being led down a particular design solution. Also, the employer may believe that more competitive bids will result (for whatever reason) by not engaging particular contractors pre-construction, which leads typically to the following contractual scenarios.

2.1.5.1.1 JCT SBC

The obligation to design rests with the employer and its design team unless it is agreed that the contractor agrees to undertake an element of design work itself. If it does, then the obligations are split insofar as the contractor is responsible for specifications contained in the Contractor's Designed Portion (CDP) insofar as they have not been provided in the employer's requirements.

If the contractor does undertake an element of design, then it must comply with any instructions from the architect to integrate its design with the rest of the works, although it can object if this would somehow have a negative impact upon such design. It does so by giving notice (seven days) to the architect who will then confirm its instruction again or indicate otherwise. Ultimately, the contractor does not have the control here and must comply if the architect disagrees.

Where not specified, the materials used for the CDP shall be of a quality appropriate to that design or of the works generally. This affords the contractor a certain degree of latitude when considering how to proceed, but may not be acceptable to an employer who wishes to assure a high level of finish to a particular project. In such circumstances, the employer would be well advised to consider specifying the complete works and getting rid of the designed portion or amending the contract with additional clarification.

One contractual tool for ensuring that the contractor has all that it needs to carry out its designed portion is the use of an information release schedule between the employer and the contractor of pre-agreed documents. This will only be successful if the employer is confident that it can deliver this information on time. Quite often, this option is not taken up as a result which can threaten the designed portion and lead to argument that the architect has not given the contractor all that it needs to discharge its obligations. This can have direct and indirect cost consequences for the employer.

In SBC the contractor is not liable for any errors or other discrepancies found in the employer's requirements. If the contractor has not picked up on any such inconsistencies in its proposals, it is entitled to a variation required for necessary compliance, but not to any amendment in the contract sum.

Notwithstanding this, if there is a discrepancy between the CDP and the employer's requirements, it is incumbent upon the contractor to set out to the architect how it intends overcoming it. The architect can either wait for that proposal or issue its own instructions if it so wishes. In any event, the removal of such discrepancy will not increase the contract sum.

The contractor's liability otherwise for design is the same as for any other independent professional designer and, whilst there is the option of a cap on its liability for poor design, that cap is rarely used as (i) it serves to sanction below standard design and (ii) it fails to consider fully the extent of remedial losses and the diminution in value of the property which an employer may suffer as a result and, consequently, not be able to recover.

2.1.5.1.2 JCT DB

This presupposes that the employer has prepared (or had prepared) certain outline requirements for which the contractor is then to take responsibility. This gives the employer a single point of responsibility for design as the contractor takes full responsibility for designs produced by others prior to its employment and, in the 2005 edition onwards (including the current 2016 edition) further responsibility for design carried out by any subcontractor.

Negotiation may ensue with the contractor unwilling to take on any design into which it has had no input prior to entering into the agreement. If that is the case, then the contract may need to be amended accordingly in a separate schedule.

In this form, the contractor takes on the design of the whole project with or without the novation of the design team.

2.1.5.1.3 JCT IC with or without design

The form is useful where the value of the works is less than for SBC and a more streamlined management process can be adopted. It supposes that the design has been worked up already and the contractor has produced a priced document in response.

2.1.5.2 NEC Suite

The core clauses state that the contractor is responsible for designing to the extent that the works information specifies the parts to be designed. If no parts are specified, then the contractor is deemed to have no design responsibilities. As a matter of best practice, it would be preferable to have that expressed in clear wording in a Z clause amendment.

If the employer is designing, it may need to appoint a design team in which case it may look to use the NEC Professional Services Contract (PSC). The benefit of doing so is that this dovetails with the Engineering and Construction Contract (ECC) but this may require amendment to cater for the inclusion of project-specific requirements, such as the provision of certain warranties to third parties.

If the contractor undertakes design work, its obligation is for its design to comply with the works information. This is, in reality, a fitness for purpose obligation. Contractors may wish to negotiate a limitation in liability to use reasonable skill and care, in which case it could consider including Option X15 which may prescribe what it is to design and how.

The risk for a contractor lies where it has not negotiated this option into its contract but engages professionals to carry out design on its behalf using the PSC. In the latter document, the professional's obligation is limited to the exercise of reasonable skill and care and therefore the contractor can be left with a greater design obligation to the one that it has subcontracted out.

Unless the contractor is going to design itself, it will need to appoint its subcontractors to do so. There is a good reason why the contractor may decide to use the NEC EC Sub-contract. If it does, the contractor is unlikely to require the project manager's consent before the appointment is made, unless there is a cost reimburseable or target element to the pricing. If the contractor opts for a more bespoke subcontract, such consent is likely to be required which may delay the process.

The important thing to note is that the NEC EC Sub-contract does not have back-to-back obligations with the ECC and therefore care needs to be taken. Its predominant use is where subcontracted design is substantial and not simply a discrete part of the works.

It may be tempting to use the short form contract and subcontract in less complex cases, but these forms are not really geared up for design control and may therefore require heavy amendment. They may be better used for construction-only contracting where the employer is clear as to what level of design it wants to achieve. Success in using the short form and its subcontract depends on careful drafting of the works information to avoid a misunderstanding of expectations. This includes consideration of project constraints which are as important as what is intended to be achieved.

2.1.5.2.1 Professional Services Contract

Of most importance to any designer is the transfer of rights option (X9) used in conjunction with the contract. This serves to transfer intellectual property rights relating to materials produced by the consultant to the employer subject to any provision in the scope.

This will be an important consideration for any employer wishing to use the consultant's design for future works. A consultant will want to limit these rights to a licence and not full transfer of ownership, unless there is some kind of cash compensation on offer for doing so.

Depending on the nature of the project, it may be possible to link the consultant's performance in design to key performance indicators and thereby provide added protection to the quality of the design produced. Option X20 is therefore a tool for doing so and would need to be incorporated into the contract.

Design considerations invariably lead to collateral warranties where third parties such as landlords, funders, tenants and purchasers are involved. Option X8 relating to third party rights may need to be incorporated if so, but it is worth knowing that the NEC does not produce a form of warranty. Invariably, parties produce their own to firm up some of the obligations expected from professional designers and prefer these to the standard forms produced by institutions in order to make them project specific and avoid heavy amendment.

Overall, the employer must spend time carefully drafting the scope, as many a dispute has been caused by parties not agreeing on what a professional has agreed to do and, conversely, what it does not wish/intend to do. The employer has flexibility in changing the scope during the course of the contract.

2.1.5.2.2 Term Service Contract

This is intended to be used for service-based, maintenance-type activities. Any design needs to have been worked up by the employer or its design team prior to entering into this contract. If a party is being asked to design, then either the ECC or EC Subcontract should be used, or the PSC.

2.1.5.2.3 Framework

The framework agreement can be operated to procure design services, although it may be that the employer has already a detailed design following an appointment of its professional team under a Professional Services Contract (referenced above). The framework does however provide for flexibility in this regard for any subsequent services to be provided.

2.1.5.3 FIDIC

This suite defines design obligations quite clearly. The Red Book is designated for employer-led design (although it also contains provision for limited contractor design in clause 4.1). The Green and Blue Books, short form contracts, are also employer-led design but may not always be appropriate save where the employer is experienced and hands on due to their brevity and potential lack of clarity.

The contractor takes responsibility for design in the Yellow, Gold and Silver Books, although there is still the requirement for the employer to outline what it wants to achieve.

2.1.5.4 ACA – PPC2000 and TPC2005 (as amended)[8]

This suite provides for a design team to be appointed. Depending on the circumstances, the contractor may or may not be part of that team.

From the client's point of view, this form gives it maximum control whilst co-ordination and co-operation of the design team co-ordinated by the lead design means that performance in terms of time and quality can be closely monitored and risk minimised. The process does require the client to play an active role.

The design process is quite prescriptive to ensure that quality and time are key aspects for its detailed production. Contained in clause 8, the form states that the design team is comprised of the parties detailed in the particulars.

The process involves the working up of outline designs and alternative solutions which are then submitted to the client, i.e. the employer. The designs are then considered by the group alongside any partnering objectives as outlined in clause 4.2 which focuses on achieving certain key performance indicators to ensure increased quality and reduced waste in terms of cost and time.

The design team require the client's approval to proceed to development of designs to the level specified in the project timetable and, at the very least, a design capable of being submitted for planning permission (if planning was ever required).

Specialist design input is then added, as required, before submitting to a further round of approvals as required by the project timetable (which

8 Discussed here only for completeness of the worked example. Please refer to other guidance for more information on the ACA forms of contract.

will state the level of design and the sequence in which approval should be sought) from the client and the core group. Working in conjunction with others in the design team, the client/core group then takes into consideration the results of any surveys and investigations provided by the partnering team. The lead designer is responsible for obtaining regulatory approvals and for overcoming any problems in relation to obtaining those approvals.

Once the works have started on site, each member of the design team is required to submit its designs for approval or for comment as stated in the project timetable and no later than the commencement of construction of each relevant part of the project. It is incumbent on the client and any other recipient to comment on any non-compliance with the partnering documents before they make any necessary adjustments.

Specific to this form, the design team must have consideration for the budget and the lead designer, with input from the others, will make amendments to achieve the necessary results following the exercise.

Another feature is the ability of the constructor, as it is called here, to object to any part of the design to the extent that it has had no role in the design itself or by virtue of any of its specialist subcontractors. This is then treated as a change and one which may give rise to adjustments in the programme.

The control afforded to the client concerning subcontractors is enhanced. There is the option for it to appoint specialist subcontractors directly if stated in the Project Partnering Agreement or 'if the Client and the Constructor agree that to do so is in the best interests of the Project' (clause 10.11). Those specialists can then be novated over if required at a later date. Obviously, to be absolutely certain, the client should foresee such a requirement at the time of contracting.

2.1.7. Summary of contracts and their uses

This part focuses exclusively on possible uses, how the contract is managed, who can be responsible for design, possible pricing options and additional high level considerations for drafting the contracts.

The table below does not consider aspects such as the inclusion of advanced payment and performance bonds or any other ancillary documents (including the relevant subcontracts and sub-subcontracts which are used with various main contract forms) or specific clauses to be drafted and incorporated into the contract. For such detailed guidance, please refer to the specific institutions' publications:

Type of contract	Uses	Areas for consideration
JCT Suite (2016 and 2011 editions)		
Standard Building Contract (SBC) with quantities	Larger works designed and/or detailed by or on behalf of the employer, where detailed contract provisions are necessary. Employer provides contractor with drawings and with bills of quantities to define the quantity and quality of the work at the time of tender.	Should ensure that all tenderers are using the same quantities prepared in accordance with the 'Measurement Rules' which replaces the 2011 definition of 'Standard Method of Measurement'. May need further clarification in the drafting. Accuracy of the quantities is the employer's responsibility, and the contract provides for the correction of errors with appropriate adjustment of the contract sum. New

	Lump sum with monthly interim payments unless otherwise stated. The contractor prices in detail the specification or the work schedules: - with the total constituting the contract sum; or - a lump sum required for carrying out the work shown on the drawings and described in the specification with an analysis of how this has been calculated with rates for valuation of variations.	provision in the 2016 edition also makes it clear that the Contractor is not only obliged to notify the Employer of a discrepancy or divergence within the Contractor's Proposals but also any discrepancy or divergence between the Contractor's Proposals and/or other Contractor's Design Documents (clause 2.16.1). The contract sum is the total arrived at in the fully priced copy of the itemised contract bills. There may be a reluctance on the part of contractors to tender without quantities where the project exceeds a certain size or complexity. The risk of error is in pricing largely by the contractor, so this may be reflected in the value of tenders submitted to cover

		preparation costs and possible disputes over the quantity of work implied in the contract documents.
SBC with approximate quantities	Larger works designed and/or detailed by or on behalf of the employer, where detailed contract provisions are necessary. Employer provides the contractor with drawings and with approximate bills of quantities to define the quantity and quality of the work, which are to be subject to remeasurement. Useful for employers where insufficient time to prepare the detailed drawings necessary for accurate bills of quantities to be produced. Price is based on the tender submitted known then contractually as the	Employer to provide at tender stage a set of drawings and approximate quantities . Possible price uncertainty for employer as tender sum is only indicative of the likely price of the works.

	ascertained final sum, subject to remeasure. Interim payments are monthly unless otherwise stated.	
SBC without quantities	For larger works designed and/or detailed by or on behalf of the employer, where detailed contract provisions are necessary. Degree of complexity is not such as to require bills of quantities. Employer provides contractor with drawings and with either a specification or work schedules to define adequately the scope and quality of the work. Price is based on lump sum with monthly interim payments unless otherwise stated.	Employer to have design completed by time of tender. Best price certainty for employer .

Design and Build (DB)	Where detailed contract provisions are necessary and employer's requirements have been prepared and provided to the contractor. Contractor completes the design as well as the works. Employer employs an agent (who may be an external consultant or employee) to administer the conditions. Price is based on a lump sum with interim stage payments or periodic payments as stated. Valuation of a change in the scope of works is based on figures in the contract sum analysis unless the Schedule 2 Supplemental Provisions apply or as otherwise agreed.	If not designing whole works, consider SBC with a design portion or the NEC or FIDIC Red Book as appropriate. The extent of the contractor's design input can vary considerably. Requires adequate time and care to compiling the employer's requirements. Desirable for changes during the works to be kept to a minimum to avoid cost increase. Employer bears cost of consultants until and unless novated to the contractor. In the event of any conflict between the employer's requirements and the contractor's proposals, the latter are stated to prevail. Contractor's design responsibility will normally be that of reasonable care and skill. The boundaries

		of design responsibility of the contractor should be carefully defined.
Intermediate	This contract is more detailed and contains more extensive control procedures than the Minor Works Building Contract (MW) but is less detailed than the Standard Building Contract (SBC). Where the proposed building works are of simple content involving the normal, recognised basic trades and skills of the industry, without building service installations of a complex nature or other complex specialist work. Design specification and quantities provided by employer for tender.	If the contractor is to design discrete part(s) of the works, consider using the Intermediate Building Contract with Contractor's Design (ICD 2016) depending on the scale of design. Conditions are less detailed than those of the Standard Building Contract. All subcontractors, whether chosen by the contractor or named by the employer, are domestic, and their performance (workmanship and materials and goods used, but not design) is the responsibility of the contractor save for certain employer obligations upon contractor insolvency in respect of subcontractors that it may have named.

	Price is based on lump sum with monthly interim payments unless otherwise stated. Contractor prices against bills given by employer or provides a lump sum with a breakdown known contractually as the Schedule of Rates or Contract Sum Analysis. Variations may be valued using this data.	Possibility of confusion regarding subcontractor design responsibility especially where design portion ill-defined.
Minor Works (MW)	Where the work involved is simple in character and designed by or on behalf of the employer. Where the employer can provide drawings and/or a specification and/or work schedules to define adequately the quantity and quality of the work:	Where the contractor is to design discrete part(s) of the works, consider using the Minor Works Building Contract with Contractor's Design (MWD 2016). Conditions much less detailed than those in the Intermediate Building Contract, and it should be noted there is no provision for naming subcontractors but subcontractors can be engaged via the JCT Short Form of Sub-Contract.

	Not suitable: - where bills of quantities are required; - where provisions are required to govern work carried out by named specialists; - where detailed control procedures are needed.	
Homeowner	A consumer contract for use by a residential occupier, drafted in clear simple language to comply with the *Unfair Terms in Consumer Contracts Regulations* 1999 (now enshrined in the Consumer Protection Act 2015). Small domestic building work such as extensions and alterations. Where the proposed works are to be carried out for an agreed lump sum.	A construction contract with a residential occupier or someone who intends to occupy the dwelling as his/her residence is excluded from the provisions of the *Housing Grants, Construction and Regeneration Act* 1996 *(as amended)* in England and Wales. Nevertheless this contract provides for adjudication in the event of a dispute between the homeowner/occupier and the contractor.

	Contract is based on drawings and/or a specification provided by the consumer. Price is based on a lump sum and the accepted quotation from the contractor. The price is to be inclusive of any VAT. There is provision for a single payment on completion, or, if agreed, payment by instalments.	After acceptance of the contractor's quotation, whatever documents are referred to in Part 1 of the contract become 'work details'. The arrangements for the work are agreed between customer and contractor and entered or signified by tick boxes prior to signing the contract. Design should be as detailed as possible to avoid changes and possible dispute with the builder.
Repair and Maintenance Contract (Commercial) (RM)	Where the work involves the repair and maintenance of a building often favoured by local authorities or employers experienced in placing jobbing work. Not suitable: - for periodic repair or maintenance over a fixed term or the regular maintenance of plant, etc.; consider using the	Flexible pricing as it enables the employer to seek quotes on the basis of a fixed price and/or of day-work or other rates, using either a Schedule of Rates or all-in labour rates in a Schedule of Hourly Charges. Express provision for payment either by a single payment or by stage payments, at the same time noting the statutory right to instalment payments

	Measured Term Contract (MTC); - for work on a dwelling by a residential occupier – consider using the building contract for a homeowner/occupier or if the work is of a very minor nature the Home Repair and Maintenance Contract (HO/RM).	that may arise under the *Housing Grants, Construction and Regeneration Act 1996 (as amended).* RM 2016 is published as a single document, comprising both the Tender (with its accompanying Invitation to Tender and form of acceptance) and the Conditions.
Management Contract	Large-scale projects requiring an early start on site. Where the employer provides the management contractor with design specification and drawings, but where it is not possible to prepare full design information before the works commence. Where much of the detail design may be of a sophisticated or innovative nature requiring proprietary systems or components designed by specialists.	The employer may need to appoint an architect/contract administrator, a quantity surveyor and such other persons as may be necessary for the professional team . Contract is divided into two periods; the pre-construction period and the post-construction period. The management contractor should be appointed early so that he/she can co-operate with the architect/contract administrator, quantity

	Where the management contractor administers the conditions; it does not carry out any construction work but manages the project for a fee. Where the management contractor enters into trade packages directly. Employer pays the management contractor the prime cost, which is paid to the various trade contractors subject to them submitting certain applications for payment and the management contractor often recommending how much is due.	surveyor and other members of the professional team on such matters as programme, formulating and agreeing construction methods, advising on 'buildability' aspects of the project, agreeing the contract cost plan, and advising on the works packages for which the works contractors will tender. After the architect/contract administrator has notified the employer that it is practicable to commence construction, the employer can then decide whether or not to proceed into the construction period. In the construction period, the management contractor will be required to set out, manage, organise, supervise and secure the carrying out and completion of the project through the works contractors, who are directly contracted to him.

		Although the management contractor is responsible for operating the terms of the contract, the consequences of any default by a works contractor do not fall upon the management contractor if he/she complies with such terms. Perennial debate as to what happens when the employer does not pay the management contractor the prime cost to pay the trade contractors. Withholding payment can be deemed to fall foul of the prohibition on conditional payments contained in the *Housing Grants, Construction Regeneration Act 1996 (as amended).* Obligations of management contractor for passing on prime cost need to be carefully considered.

Measured Term	Where the work is to be instructed from time to time and measured and valued on the basis of an agreed schedule of rates. It does not matter that large number of orders may be placed with the contractor over the period of the contract and that they include orders of widely differing values and cover various different works. Based on a priced schedule of rates to be followed by a series of orders for each separate item of work. Where the contractor is to design discrete part(s) of the works (contractor's designed portion). Order may include a written description and drawings where relevant.	Requires forethought by the employer to: - list the properties to be covered by the contract and state the period and the type of work which may be required from time to time during that period (the period is envisaged as being not less than 12 months and it is seldom safe or practical for it to exceed 36 months); - estimate the total annual value of the work for the whole contract period, and indicate the minimum and maximum value of any one order given; - appoint a contract administrator who will issue the orders from time to time describing the work to be completed under each order and the completion date; and - pay for each item of work covered by an order, as certified by the contract adminis-

	Price is based on measurement and valuation for each separate order, using the figures in the priced schedule of rates as adjusted by the contractor and accepted by the employer.	trator, following measurement and valuation according to the relevant rates and prices in the priced schedule of rates, which are to be adjusted by applying the contractor's quoted percentage addition or deduction.
Prime Cost	The work proceeds on the basis of a brief specification, drawings (if any) and an estimate of its cost. Projects requiring an early start on site, where the works are designed by or on behalf of the employer but where it is not possible to prepare full design information before the works commence. Where detailed contract provisions are necessary and the employer is to provide a specification describing and showing the items of work.	High risk for the employer in terms of cost, and much depends on the efficiency of the contractor in carrying out the works economically. Various provisions to assist in keeping the expenditure of prime cost to the minimum needed to provide the works required by the completion date. The contract fee can be revised if the employer changes the nature and scope of the works described in a schedule to the contract. After the contract has been entered into, any such

	Where an architect/contract administrator and quantity surveyor are to administer the conditions. Interim payments are monthly unless stated otherwise. The contractor is paid the prime cost of the works, as certified by the architect/contract administrator. The employer also pays a contract fee in respect of the contractor's non-site overheads and profit. This fee may be a lump sum which can nevertheless be adjusted if the actual prime cost is more or less than the estimated prime cost by a percentage stated in the contract, or it can be a 'percentage fee' calculated on the actual prime cost incurred.	change is also subject to a right of reasonable objection by the contractor

Pre-Construction Services agreement	Interim appointment of a general contractor by the employer to carry out pre-construction services, under a two-stage tendering procedure. The appointment follows first stage tenders and covers the period leading up to the contractor's submission of a definitive second stage tender and entry into the main contract for the construction phase. There is a separate version for the appointment of a specialist for carrying out pre-construction services. Where the main contract is to be the JCT Standard Building Contract, Design and Build Contract, Major Project Construction Contract, Intermediate Building Contract or Interme-	Enables the contractor to assist the consultant team with the development of detailed designs and development of the main contract works and specialist tender documents. The contractor's involvement at pre-construction stage is widely viewed as being valuable and often essential in the final design process and preparations for the construction phase, including the programme, cost plans, buildability and specialist procurement. The agreement can be used whether or not the contractor is to be responsible for design work but, unless otherwise agreed in the provision of services, any liability will only arise once a contract for the construction works is executed.

	diate Building Contract with Contractor's Design, 2016 edition. Whether or not the contractor is to be responsible for any design work. Where there is to be novation to the contractor of any specialist subcontract(s) or supply contract(s) or (in the case of a Design and Build Contract or Major Project Construction Contract) any consultancy agreement(s). In a JCT Construction Management procurement, with minor amendment, for the provision of preconstruction services by prospective trade contractors. Not suitable for use: - between the employer and specialist subcon-	

	tractors (except as prospective trade contractors in a JCT Construction Management procurement); - between a contractor and a subcontractor; or - in conjunction with the JCT Management Building Contract.	
Constructing Excellence	For the procurement of construction works and construction related services. For use throughout the supply chain including the provision of professional services. For use where participants wish to engender collaborative and integrative working. For use in partnering. Whether or not the supplier is to design.	Encourages collaborative behaviour and expressly underpins collaborative working and the formation of integrated teams. It provides specifically for the use of a risk register, risk allocation schedules and performance indicators. Simply worded but lacks detail and possible enforceability. The extent of the supplier's design input, which may be as a consultant or as a contractor, can vary considerably.

		Can lack certainty in terms of decision making.
Major Projects	For major works where the employer regularly procures large-scale construction work. Where the contractor to be appointed is experienced and able to take greater risk than would arise under other JCT contracts. Where the parties have their own detailed procedures and where limited procedures only need to be set out in the contract conditions. Where the employer has prepared his/her requirements and provided these to the contractor. Where the contractor is not only to carry out and complete the works, but also to complete the design.	The contractor assumes more risks and responsibilities than under other JCT contracts. It is desirable, therefore, that the employer and the contractor, together with their respective advisers and subcontractors, are experienced in detailed risk management and undertaking large commercial projects.

	The employer employs a representative to exercise the powers and functions of the employer under the contract. Price is based on lump sum with monthly interim payments based on the pricing document. This contract is for use on major projects where the employer and the contractor regularly undertake such projects and have appropriate in-house contractual procedures.	
Framework	For the procurement of construction/engineering related works over a period of time. For use by clients with contractors and/or suppliers. For use by contractors, subcontractors and/or suppliers sub-letting	Employers who carry out work regularly and wish to try and capture the benefits of long-term relationships within the supply chain should consider the use of the JCT Framework Agreement in conjunction with the appropriate JCT contract for the particular project.

	to others in the supply chain. For use with most standard forms of construction and engineering contracts and subcontracts. Where compliance with the public procurement rules is required. On a single project.	
Construction Management	The appointment is suitable where: - a construction manager is to manage the project on behalf of the employer; - the employer is to enter into direct separate trade contracts using the Construction Management Trade Contract (CM/TC) or a special Trade Contract.	This appointment is the 'main contract' for the procurement path of construction management. It is part of a suite of documentation for use where separate contractual responsibility for the management, design and construction of the project is appropriate. It is drafted as a professional appointment for a construction manager.

	The accompanying trade contract is suitable where: - the employer is to enter into direct separate trade contracts; - a construction manager engaged under the construction management appointment is to administer the conditions on behalf of the employer. Price is based on either a lump sum or complete remeasurement. Interim payments are either monthly or at predetermined stages.	The accompanying trade contract is not a subcontract but more like a main contract between the employer and trade contractors. Under it, the construction manager acts as agent for the employer in issuing instructions, making decisions and preparing certifications. The employer, and not the construction manager, remains liable for payment.

Form	Use	Areas for consideration
NEC3 Suite		
Option A – lump sum	Definite scope of works against which the contractor prices with activities.	Risk to cash flow if not priced evenly as contractor only gets paid once activity completed.
Option B – remeasurement	Can include lump sums, quantity and time-based items. Where risk of quantities changing is quite high, as in groundworks.	If the quantity is not correct when the work is carried out, then it is corrected and paid for work actually carried out. This minimises the impact of getting it wrong for the contractor. Don't have to wait until an activity has been completed to get paid, as payment is worked out on a quantity basis.
Option C – target cost with activity schedule	Used when high chance of quantities changing. Contractor works to a target cost schedule and also charges a fee in respect of its works and any subcontracted works.	Can be administratively burdensome. Strong incentive on contractor, who takes on increased proportion of risk, to not exceed the target.

	Contractor paid his/her actual cost and his/her fee throughout the project. At completion, the cost is assessed. If there are cost savings, these are split in the proportions agreed in advance with the employer. If the costs are exceeded, then these costs are due to be paid back. Calculated again at final account stage using final figures.	
Option D – target cost with bill of quantities	Actual cost established by bill of quantities not activity schedule. Target price is changed to allow for changes in quantities and compensation events.	Preferable over C if possible . Higher risk for employer.
Option E – cost reimbursable	Urgent works or where work cannot easily be assessed from the outset.	Almost zero risk to contractor. May sanction incompetence.

Management fee	Same as management contracting under JCT.	More expensive for employer but due to management contractor taking on contractual risk.

Form	Use	Preparation points
FIDIC Suite		
Red Book – The Conditions of Contract for Construction	Building and engineering works designed predominantly by the employer.	All subject to Autumn 2017 revisions but, currently: May be wise to include only the general conditions with relevant project documents incorporated in a particular contract, as the accompanying guidance contained within the contract may simply cause unnecessary confusion. The conditions are not stand-alone, so require additional drafting and co-ordination to ensure that they serve their intended purpose such as: - ensuring that applicable laws tally with the contractual law applicable; - any requirements of any financing party;

		- any project-specific procedures including any pre-construction consents. Suitable for all projects where main responsibility for design lies with employer (or his/her engineer) Any design portion undertaken by the constructor may need to be defined very carefully. Contains extensive project management provisions by engineer designed to avoid disputes from arising such as giving approvals for work, certifying payment, measuring work according to a Bill of Quantities or in accordance with a lump sum.
Yellow Book – Conditions of Contract for Plant and Design-Build	Building and engineering works designed by the constructor where testing and commissioning is more important such as manufactured plant, turbines, generators, switchyard equipment.	Design element not remeasured to avoid uneconomic design by the constructor. Some design may be done by the employer. Design performance related subject to testing after completion. Risk

	Suitable for all types of projects where main responsibility for design lies with the constructor. Recommended for the provision of electrical and/or mechanical plant, *and* for the design and execution of building or engineering works. A two-stage tendering process may be appropriate.	that various performance criteria not tied in contractually or missed due to preparation of criteria by multiple consultants in different documents. Still requires the employer to prepare employer's requirements and to assess tenders, which can be an onerous exercise. Obligation on employer to evidence to constructor that it has the finance to pay. Consider the payment guarantee at Annex G to satisfy this requirement. Payment terms usually on a lump sum basis, according to milestones achieved, e.g. a certain percentage when manufacture and testing at the factory is complete, after shipping and delivery to site, when erection is complete and on satisfactory commissioning. Administration of contract and supervision by engineer for approvals certifying payment.

		Payment on lump sum basis, usually against a Schedule of Payments. Testing procedures usually more complicated than for 'Red Book' and performance may be assured by entering into a type of on-demand bond within a fairly stringent time frame.
Silver Book – Conditions of Contract for EPC/Turnkey contracts	Constructor takes total responsibility for design and completion of the project, ready for occupation. Use when employer wants greater certainty of price and time and is prepared to pay for it. A two-stage tendering process may be appropriate. Suitable for electrical and mechanical and other process plant projects for: - all types of employers;	More expensive (employer may even pay the tenderers for the second stage), more time-consuming pre-contract to negotiate terms. Employer may not get best value if it wants to be hands-on during the construction or engineering process. Responsibility for design lies solely with constructor. Employer provides 'Employer's Requirements' to which constructor designs.

	- often in civil law countries; - where government employer or private developer wants his/her project on a fixed-price turnkey basis and with strictly two-party approach (i.e. no engineer). Not suitable: - where high levels of unforeseen risk; - if time or information is insufficient before contract signature; - if considerable work is underground or difficult to inspect; - if employer intends to supervise closely or control or review; - if an intermediary determines interim payments; - where part of works is designed by employer;	Employer's Requirements usually 'performance specification' type. Constructor carries out all engineering, procurement and construction, providing a fully-equipped facility, ready for operation at 'the turn of a key' No engineer – instead it is the employer who may appoint an employer's representative. Lump sum contract price extended testing procedures including tests after completion. Constructor takes majority of risk, employer pays more to cover such risk. Final price and time of completion should be more certain.

	- for public bidding without negotiations.	
Green Book – Short Form of Contract	Building and engineering works of relatively small capital value, short time period or relatively simple works. For use by experienced FIDIC parties.	May need to negotiate an agreed solution as brevity means that procedures may not be covered.
Blue Book – Form of Contract for Dredging and Reclamation Works	Particular to these activities.	
MDB (Multilateral Development Banks) Harmonised Construction Contract	Currently, only used with participating funders but could change after Autumn 2017.	Since funding is dependent upon agreement to the standard form, there may be little room and resistance to amendment even for project-specific reasons.

2.2
ACHIEVING THE CORE OBJECTIVES: WHICH CLAUSES ARE AVAILABLE?

Having ascertained which potential contracts may fit your client's needs, you now need to consider which clauses within each are going to be most appropriate to match those aims and secure its interests best.

Below is a general run through of the relevant provisions you will come across with a brief explanation as to how they interact and some of the ways in which they compare and contrast between each standard form.

By focusing in on what is important to your client's goals, this should help you craft provisions within a particular contract which reflect more accurately on paper the types of issues which need to be managed in reality on site.

2.2.1 Pricing, payment and funding: cost

Summary of payment options				
Contract	**JCT**	**NEC**	**FIDIC**	**PPC2000**
Lump sum fixed price contracts	Yes. DBYes. SBC, IC and MW Forms.	Options A and B.	Yes. Red Book, Yellow Book and Short Form.	No. But 'agreed maximum price' available.
Remeasurement contracts	Yes. SBCYes. and IC.	Option B.	Yes. Red Book & Short Form.	No.

Target cost contracts	No. But the Yes. constructing excellence contract can be used as a 'target contract'.	Options C and D.	No.	Yes. By reference to key performance indicators and shared savings to be considered.
Cost plus options	Yes. Yes. Management contract and prime cost contract.	Options E Yes. and F.	Short Form.	
Additional charges	By way of variations, day works, fluctuations, and loss and expense.	By way of compensation events.	Adjusted for scope changes and other adjustment factors (as applicable) as set out in clause 13.	Cost plus, open-book approach.
Interim payments	Yes. Allowed for in all forms.	Yes. Allowed for in all options.	Yes. Red Yes. Book, Yellow Book and Short Form.	

Lump sum fixed price contracts have historically been favoured by funding institutions, due to their cost certainty. However, fixed price procurement will not suit every project, particularly if there are uncertainties as to scope and/or other significant variables.

The contract suites offer a number of pricing options to their users, of which some are used more heavily than others. Commercial and domestic projects in the UK must all contain Construction Act compliant (Housing Grants Construction and Regeneration Act 1996 as amended by the Local Democracy, Economic Development and Construction Act 2009) payment terms and these requirements are reflected in the terms of most of the standard forms other than FIDIC.

2.2.1.1 JCT

The JCT's comprehensive suite of contracts caters for most pricing options, although currently no dedicated 'target cost' cost form has been published (the Constructing Excellence Contract can be used on a target cost basis).

This leaves the DB (Design and Build) and SBC (Standard Building Contract) contracts as the key lump sum options. The contract sum in the DB form is based on a contract sum analysis. The SBC sum is priced either on the basis of specifications/works schedules or on the basis of a contract sum analysis/schedule of rates.

The priced document is the equivalent of the contract sum analysis in the SBC with quantities edition. The SBC with quantities contracts assume that a fully priced bill of quantities will be used. The contract also gives the parties an option to work with a priced schedule of activities (the schedule will not cover provisional sums or work items where approximate quantities have been established). Adjustments in the contract sum may be by way of addition or deduction, and adjustments for inflation are optional.

Section 4 in each case deals with payment. There is an advance payment option, implemented by an appropriate entry in the contract particulars. Whilst the contract assumes that the employer will operate a retention mechanism, the use of a retention bond in lieu of retention deductions is optional. Where retention deductions are made, half the retention is released at practical completion and the balance is released after the certificate of making good is issued. Amounts for goods and materials stored off-site can be included in interim payments subject to certain conditions, which may involve the provision of a bond. When produced, and unless formally disputed within the stipulated time period, the final certificate becomes conclusive as to the balance due between the parties.

Additions to the contract sum fixed price arrangements can only be claimed if the contractor can demonstrate that there has been a change (under the DB) or a variation (under the SBC) or that it has incurred loss and/or expense as a result of a relevant matter, or that an adjustment to the contract sum is otherwise due.

The intermediate contracts are generally also fixed price agreements (there may, however, be some re-measurement where approximate quantities are included in the Bills of Quantities) priced on the basis of bills of quantities/specifications/works schedules or a contract sum analysis. The contract sum may only be increased where there has been a variation or where the contractor has incurred loss and expense.

Interim payments are made and can include amounts for goods and material stored offsite, provided that the relevant conditions are satisfied. The default position for inflation is that the fluctuation options will apply unless expressly excluded in the contract particulars. As might be expected, parties have a shorter period within which to settle accounts for the final payment.

In the Minor Works contracts, works are priced on the basis of a schedule of rates or a priced specification/work schedules. This contract form is not suitable for use with a bill of quantities. Progress payments of a percentage (stated in the contact particulars) of the total value of properly executed work, and of goods and material brought to site, are made at intervals of four weeks. Interim payments on and after practical completion are also provided for. Unless the fluctuation option is chosen, the contract is let for a fully fixed price. There is no contractual entitlement to payment for loss and expense, simply a general right to request extensions of time for 'reasons beyond the control of the contractor'. All additional charges are to be incorporated into the variation account.

2.2.1.2 NEC3

Options A (priced with activity schedule) and B (priced with bill of quantities) are contracts based on lump sum arrangements. The remaining options are based on cost reimbursement principles. Of the latter, Options C and D are 'target cost' contracts[1]. Options E (cost reimbursable contract) and F (management contract) are 'cost plus fee' options. Whilst the works contracts let under Option F may themselves be lump sum, the management contractor is paid the costs of such packages plus a fee. Under Options C and D, the subcontracts may be procured on a lump sum (or any alternative) basis, notwithstanding the target cost arrangement in the principal contract. The NEC publishes equivalent forms of subcontract compatible with the principal contracts.

In the fixed price contracts, the price is set out in the activity schedule or the bill of quantities. The contractor is paid the price and any additional payments are determined by the compensation event mechanism. Under the other options, the contractor is paid according to the defined costs incurred for the work done to date (including additional costs, whether or not due to compensation events) less any disallowed cost.

In each case the project manager certifies payment and the contract determines the frequency of interim payments. Construction Act compliant payment clauses for use in the UK can be incorporated, as necessary, using secondary option clause Y(UK)2. The secondary option clauses also allow the parties to agree and use additional pricing provisions and financial incentives (for: inflation adjustments (X1), payment in multiple currencies (X3), payment of an early completion bonus (X6), advanced payment (X14), retention (X16) and key performance indicators (X17)).

1 Neither form uses the expression target cost but they are called 'target' contracts. The target is the total of the lump sum prices in the contract (adjusted by compensation events) which, compared with the defined costs incurred on the project, determines the savings or overspend against target: the 'pain/gain' share.

Additional costs under Options A and B are recovered in a similar way to additional costs under a JCT contract. The contractor must demonstrate either that there has been a change in the works information or that costs relate to another compensation event. The approach under the other options is, by contrast, that all defined costs are reimbursed provided that they are admissible, and in the case of Options C and D subject to the application of the 'pain/gain' mechanism. The employer must establish which costs constitute disallowed costs. Unlike the fully cost reimbursable options, Options C and D require the contractor to use very carefully the early warning and compensation event mechanisms to see that target cost is adjusted properly.

2.2.1.3 FIDIC

The Red Book's guidance notes for the preparation of particular conditions highlights various price and payment issues for consideration. The Red Book may be used as either a re-measurement (work done is measured and payment is made in accordance with the bill of quantities) or a lump sum contract. If lump sum, clause 12 (Measurement and Evaluation) should be deleted and amendments made to clauses 13.3 and 14.1. The notes also suggest amendments that might be made to the contract for a cost plus arrangement, however the cost plus route is not widely endorsed due to the inherent risks for the employer.

The contractor is paid the contract price (clause 14.1) which is adjusted for scope changes and other adjustment factors (as applicable) set out in clause 13, employer's risks, and force majeure events. Advance payments are possible (clause 14.2) and the contractor may apply for interim payments (clause 14.3) which are paid on issue of a certificate (clause 14.6). The Appendix to tender should specify the total amount of advance payments and the basis of repayment.

The Yellow Book is intended to be a lump sum contract, consistent with the approach to design and build and 'turnkey' options in other standard forms. There is accordingly no equivalent to clause 12 of the Red Book in this contract and additional drafting will be needed if the project requires that certain works elements are to be re-measured.

With little or no re-measurement in this contract, the contractor takes the risk of any cost adjustments due to his/her design. The contract assumes that interim valuations will be based on a schedule of payments (clause 14.4). Any alternative method of determining interim valuations must therefore be stated in the particular conditions.

The Short Form of Contract can be used for all pricing options, from lump sum through to full cost reimbursement. Potentially therefore a hybrid pricing arrangement could be operated under Short Form if carefully defined. The method of valuation is selected by completing the relevant entry in the Appendix.

Payments are made monthly upon presentation of the contractor's statement (clause 11.2). Retention may be deducted, together with any amount with which the employer does not agree. Interim payments maybe be wholly withheld until the employer receives the required performance security (clause 11.3). The contract assumes that payments will be made in a single currency, therefore if multiple currencies are to be used, the proportions of each payment to be made in a particular currency will need to be stated in the Appendix and particular conditions added. The Short Form is intended for use in short duration projects, therefore for works of longer duration, provisions will need to be adapted to manage the effects of inflation.

In addition, the payment mechanism in the FIDIC suite is structured for international projects and therefore not UK Construction Act compliant. Amendments to the payment provisions will therefore be necessary if either contract is being used for a UK scheme or an international project governed by English law.

2.2.4 Change control: quality and cost

2.2.4.1 JCT

In the SBC, the architect/contract administrator has the power to order variations (clause 3.14), which includes verifying variations requested by the contractor. Variations are contained in section 5 and are classified broadly as:

- an addition, omission or substitution of any work (it should be noted though that, as at common law and in many other construction contracts, work cannot be omitted so that it can be given to another contractor);
- the alteration of contractual standards for the materials or goods;
- the removal from site of work executed or materials that are in accordance with contract;
- any alteration of working restrictions such as access to the site, site hours, working areas or the order of work sequencing; or
- the expenditure of provisional sums.

One of the most notable of these is the power to issue instructions to address working restrictions onsite. Where there is a new build project on unoccupied land this may not be much of an issue, but, for example, where the contractor is carrying out refurbishment works in an operational building, the contractor's ability to do the work to a satisfactory standard could be adversely affected if variations of this nature could not be instructed.

Variations are to be valued either:

- by agreement between the employer and contractor;
- if agreement cannot be reached, by a valuation of the quantity surveyor, in accordance with the valuation rules; or
- by the architect/contract administrator requesting and the contractor issuing a variation quotation (clause 5.3 and Schedule 2).

It has been held that a valuation made by the quantity surveyor is only a recommendation to the architect/contract administrator, who is not bound to accept it[2]. If a variation quotation is requested, then the contractor does have a right of objection. If the contractor agrees, then the procedure to follow is set out in Schedule 2 of the contract.

Otherwise, the valuation rules are relatively consistent between the forms. Where a variation is similar to the contract works, then it is valued in accordance with any rates stated in the contract. If not, then it is valued at 'fair rates and prices'. An appropriate addition or reduction shall be made in terms of preliminaries relating to the variation. If for some reason a variation cannot be priced in this way (clause 5.6 or 5.8, as relevant), then it should be valued at prime cost plus percentage additions (as stated in the contract particulars).

A variation also entitles the contractor to an extension of time (clauses 2.29).

In the DB form, the term 'change' is used instead of 'variation', but the definition is broadly similar. The DB form makes clear that a change will only relate to a change in the employer's requirements. The valuation rules are slightly different to the SBC.

The primary position is that agreement should be reached between the employer and the contractor, as above. However as the DB form does not include the role of a quantity surveyor, the valuation rules (clauses 5.4–5.7) are operated if there is no agreement. The rules in this case expressly allow for additions for increased or decreased design work.

Again, if for some reason a change cannot be assessed in this way, then it should be valued at prime cost plus percentage additions (as referenced in the contract particulars). A change also entitles the contractor to extra time (clauses 2.23–2.26). Provisional sums are valued as a change.

The differences between the intermediate forms and the SBC are that the intermediate forms make allowances for approximate quantities and do not make provision for a contractor's quotation (clause 5.3 of the

2 RB Burden Ltd v Swansea Corporation [1957] 1 WLR 1167

SBC). There is a slight difference in the intermediate valuation rules, than the SBC equivalent:

- if the work under the variation is of a similar character, carried out under similar conditions, and is of a similar quantity then rates and prices in the contract will be used;

- if only the condition and/or quantity is sufficiently different then the rates and prices in the contract are used as the basis, but a 'fair allowance' is given for the differences; and

- if the character of the work is dissimilar, regardless of condition or quantity, then fair rates and prices are used.

- if the variation relates to an approximate quantity, then:

    - If the approximate quantity is reasonably accurate, then the rate or price of the approximate quantity is used;

    - if the approximate quantity is not reasonably accurate then the rate or price will still be used, but shall also include a fair allowance for the difference in quantity.

The valuation is not to include an amount for direct loss and expense, but a variation does entitle the contractor to extra time (clauses 2.19–2.20).

In the Minor Works forms, the provisions are, as one would expect, considerably simplified (clause 3.6):

- the architect/contract administrator is entitled to issue instructions varying the works;

- the architect/contract administrator and the contractor shall attempt to agree an amount, increase or reduction, for the variation; and

- if agreement cannot be reached then the architect/contract administrator shall make a fair and reasonable assessment using any relevant prices in the contract, and the assessment shall include loss and expense incurred by the contractor for any delay caused.

If provisional sums are instructed, then these shall be assessed in accordance with clause 3.6. The contractor will be entitled to an extension of time for a variation provided that the need to vary does not arise from a contractor fault (clause 2.8).

2.2.4.2 NEC3

The apparatus for dealing with a variation is included in the compensation event provisions (section 6, clauses 60–65). Compensation events are based on similar events or circumstances as relevant events and relevant matters in the JCT suite.

Eight of the events are based on instructions, communications or certificates issued or neglected by or on behalf of the employer; there are five events relating to acts of prevention of the employer; two relating to inspections or testing; the balance are force majeure type events, unexpected site conditions, adverse weather conditions, and the occurrence of any employer's risks.

If the project manager notifies the contractor of a compensation event which requires additional work by the contractor, then the contractor is to provide a quotation. The contractor gives notice of a compensation event which has happened or which s/he expects to happen. If he/she does not do so within eight weeks, he/she will not be entitled to a change in the prices (the individual lump sums for each activity on the activity schedule) or the completion date. This is a condition precedent to the contractor's entitlement to claim, and should be borne in mind by both parties when operating these provisions.

If the contractor fails to give notice within the time period, s/he will only be entitled to a change in the prices or the completion date if the project manager should have notified the contractor of a compensation event but did not. Therefore whilst the events and the notification obligations are not dissimilar to the JCT regime, the philosophy is quite distinct, with a greater emphasis on prompt notification and proactive risk management.

The contractor supplies his/her quotation, which can be accepted or revised; a decision which is up to the project manager. There is a short timetable for the contractor to submit the quotation and for the project manager to respond one of four ways; either he/she may:

- instruct the contractor to issue a revised quotation;
- accept a quotation;
- state that he/she will no longer be making an instruction or change;
- state thats/he will be making his/her own assessment under clause 64.

The assessment of the change in the prices is broken down into the effect of the compensation event on the defined cost of work performed, the forecast defined cost of the works still to be carried out, and the fee. The assessment will not only include additional money (or a reduction for an omission) but will also include additional time (or a reduction for an omission).

The NEC3 Short Form provisions are slightly simplified from the full NEC3 engineering contract. The 19 compensation events in the long form have been reduced to 14, but the principles are the same as the long form. The provisions for notifying a compensation event, and submitting a quotation are also broadly similar.

The assessment of a compensation is confined to the impact on the defined cost, as there is not a defined fee under the contract. There is however a percentage added for overheads and profit to the defined cost for people and a percentage for other defined cost, which would therefore represent an increased amount.

2.2.4.3 FIDIC

The Red Book (clause 12 (Measurement and Evaluation); clause 13 (Variations and Adjustments)) uses the term 'variation', which is defined as 'any change to the Works, which is instructed or approved as a variation under Clause 13' (clause 1.1.6.9). A more detailed list of what can and cannot constitute a variation is set out later in the contract (clause 13.2) such as:

- changes in quantities or quality;
- the levels/position/dimensions of the works;
- omissions or additional work, plant, materials or services, which includes additional testing;
- changes to the sequencing or timing.

A variation can also include expenditure of a provisional sum and can be initiated by either an engineer's instruction, or a request by the contractor to submit a proposal. The contractor has a very limited right to object to an instruction, (such as not being able to acquire the goods) but in any event the engineer can simply ignore him/her and confirm the instruction.

The engineer also has a discretion to request a proposal from the contractor prior to issuing a variation. The engineer can however accept, reject or comment on the proposal, with no time limit under which to do so. There is no provision to cover how many sets of comments go back and forth, but the contractor is expressly prevented from delaying any work whilst waiting for a response.

The conditions specifically provide for value engineering, which, in certain circumstances, allows for a fee to the contractor. The contractor can make a proposal to do so at any time, for the purpose of shortening the contract programme, reducing the cost, increasing efficiency of the works or to 'otherwise be of benefit to the Employer'.

The FIDIC valuation rules (clause 12) are similar to those in the JCT. For example, for similar work to the contract works, rates and prices in the contract are used. New rates are used if certain thresholds are exceeded, for example, the quantity increases by more than 10 per cent, or the work is not similar to the contract works. Until a new rate/price is determined, the engineer can fix an interim rate/price for the purpose of interim certificates.

Variations can be valued on a day works basis (clause 13.6), but only if the work is of a minor or incidental nature and a day work schedule is included in the contract.

In the Yellow Book 'variation' is defined as 'any change to the Works, which is instructed or approved as a variation under Clause 13' (clause 1.1.6.9). There is not the same list of potential variations as in the Red Book, but it can be assumed that the same examples apply. The value engineering provisions in the Red Book state that if there is a change in design, the default position is that the contractor is to design the change. These provisions are not however included in the Yellow Book because the contractor has primary design responsibility for the whole of the project.

A variation is defined in the FIDIC Short Form in a more simplified manner than in the Red Book. It is a change to the specification and/or drawings as instructed by the employer. The employer has the right to instruct a variation and the contract prescribes the method of valuation (clause 10). Within 28 days from the instruction of the variation, the contractor is to issue an itemised assessment of the cost of the variation. This is not expressly stated to be a condition precedent, but an employer may wish to argue that it should operate in that way or amend it to have that effect to avoid any ambiguity.

The parties should attempt to agree the cost of the variation, but if this is not possible then it will be up to the employer to make an assessment. Once this assessment is made then the contractor appears to have no right of appeal apart from formal dispute resolution procedures.

The assessment is based on similar principles to the other construction contracts:

- at a lump sum to be agreed;
- at appropriate rates in the contract;
- at appropriate new rates which may be agreed or the employer deems appropriate. This gives the employer a highly discretionary position; or
- if the employer instructs, at day work rates.

2.2.5 Insurance: cost and time

Adequate insurance is a key risk management tool for any project. The standard forms vary quite considerably in their treatment of insurance, so it is essential that the parties understand what their chosen contract requires of them in this context and whether or not additional or alternative insurance arrangements need to be made. Insurance clauses, despite their importance, are some of the most widely misunderstood provisions in construction and engineering contracts.

2.2.5.1 JCT DB and SBC (Section 6, clauses 6.4–6.12)

The JCT insurance provisions are the regime with which most domestic construction professionals and advisers will be familiar. This is due largely to repeated and widespread use of the JCT, but also because the regime is standardised throughout the suite and less frequently amended than other JCT clauses (major procurers in both the private and public sectors are more likely to amend the insurance clauses to reflect their preferred insurance solutions, which may not align with the JCT). This is in contrast to the NEC3 or the PPC2000, under which the parties choose the relevant insurances and who is to procure them, on a project-specific basis.

On the one hand, the JCT regime is less flexible than others, but on the other the parties operating the contract may be more aware where the risk lies. It is often a major problem on building projects that the parties think the contract states one thing as regards insurance, when in actual fact it is the opposite. The JCT guidance notes themselves recognise the importance of ensuring that the words used in the contract reflect the policy arrangements, suggesting that changes to the standard terms may be required to bring the two into line.

The basic JCT regime requires the contractor to take out:

- employers liability and public liability insurance covering both injury and death to any person, and injury and damage to property;

- insurance in the joint names of the contractor and the employer, to cover injury or damage to property due to collapse, subsidence, vibrations, weakening or removal of support or lowering of groundwater attributable to the carrying out of the works. The contractor is only responsible for damage to property due to negligence, breach of statutory duty, omissions or default on his/her part;

- all risks works insurance in joint names. The three options (A, B and C) are discussed below;

- if the contractor has design responsibility, professional indemnity insurance will be required for the limits stated in the contract particulars.

Under the JCT, one of three insurance options applies, A, B or C. The details of these provisions are contained in Schedule 3:

- Option A applies to new buildings and is an all risks policy to be taken out by the contractor;

- Option B also applies to new buildings and is an all risks policy to be taken out by the employer; and

- Option C applies to extensions/alterations to existing buildings, and is to be taken out by the employer. The works are to be insured under an all risks policy, whereas the existing buildings are only to be insured against the specified perils (fire, lightening, explosions, and so on).

The all risks insurance options specify that the joint names policy should either name each subcontractor or include a waiver by the relevant insurer of any right of subrogation which the insurer may have against any subcontractor (clause 6.9) in respect of loss and damage to the works caused by specified perils.

There are certain contractual and practical issues that arise between the different options. For example, under Option B, if loss or damage is sustained by a risk covered under the all risks insurance, the contractor is entitled to a variation for the cost of the reinstatement. Under Option A, the contractor can only use the money received from the insurance policy for the cost of reinstatement, and the contractor must cover any shortfall. In addition, the contractor will only be entitled to an extension of time under Option A if the risk was a specified peril.

There are also provisions relating to terrorism cover. Managing terrorism risk is becoming more prevalent in construction contracts. Terrorism cover in the UK is provided by many mainstream insurance companies, but it is backed by the 'Pool Re' scheme. Pool Reinsurance Company was formed as a joint enterprise between the insurance industry and the UK government in 1993, following terrorist attacks related to the then situation in Northern Ireland. Prior to the World Trade Centre attacks in 2001, terrorism cover was only available for terrorism caused by explosions or fire. Since then, the Pool Re scheme has expanded to an 'all risks' basis, to cover other types of terrorism. An insured party can obtain an extension to an existing insurance policy to cover this risk through its usual insurance broker or provider.

In the event that terrorism cover is no longer available, the employer can opt whether to determine the contract, or continue. If the employer opts to continue and the works are damaged by terrorism then the contractor is obliged to reinstate the damage, but it will be treated as a variation.

Unlike the NEC3 there is no room in the contract particulars to enter in details of further insurances, and as a result specific contract amendments must be made if further insurances are required.

The clauses (Section 6, clauses 6.4–6.12) in the DB form are very similar to the SBC, except that where under the SBC the contractor is only obliged to obtain professional indemnity insurance in respect of the contractor's designed portion, under the DB form the contractor is obliged to obtain insurance for the whole of the works.

2.2.5.2 JCT, IC, ICD and MW

Again, the insurance provisions (Section 6, clauses 6.4–6.10 and 6.15–6.16). are very similar to the SBC. However, as named subcontractors are provided for in the intermediate forms, where the employer is required to take out insurance Option C, the named subcontractors are also required to be named on the insurance policy (clause 6.9.3).

The insurance regime in the Minor Works forms is slightly different to the other, more lengthy JCT forms. In an attempt to simplify the contract, the insurance provisions are solely contained in the body of the conditions, rather than the schedules. The contractor is still required to maintain employers' liability and public liability insurance, but there is no equivalent of clause 6.5.1 insurance, nor is the contractor required to maintain professional indemnity insurance, even where responsible for design under MWD.

The Minor Works forms still have three options for insurance of the works:

- the contractor takes out and maintains an all risk policy in the joint names of the contractor and the employer for the full reinstatement of the works (clause 5.4A);
- the employer takes out and maintains a joint names policy for any existing structures, including the contents, for loss or damage caused by a specified peril; and an all risk policy for the full reinstatement value of the works (clause 5.4B); and

- the employer in its own name takes out and maintains, in respect of the existing structures, and the contents for which it owns or is responsible, insurance for loss or damage caused by a specified peril (clause 5.4C).

If any damage or loss is caused to the works (under clause 5.4A), the contractor authorises the payment of any insurance monies to be made to the employer. If the contractor is required to remedy the loss or damage, the contractor is only entitled to any payment if an amount is received under the insurance policy. However, the employer can deduct from that amount professional fees that it has incurred in relation to the loss or damage.

Under the public liability indemnity (clause 5.2), where clause 5.4B applies, the contractor is not responsible to the employer for loss or injury to third party property or third parties caused by a specified peril. The employer is to procure insurance for a loss or damage caused by a specified peril, but only for the reinstatement, repair or replacement value. Therefore, where clause 5.4B applies, the employer may lose its right to claim liquidated damages for all or part of any delay in completion of the works, which is caused by a specified peril. This is because a specified peril can include events or circumstances caused by contractor default, but this would not entitle the contractor to an extension of time. If any loss or damage is caused which falls under the all risks policy (under clause 5.4B), the remedial work the contractor is required to perform will be instructed as a variation. The contractor may be entitled to an extension of time, as long as the loss or damage caused to the works was not the fault of the contractor.

Clause 5.4C is specific to the Minor Works contract. It is only in respect of the existing structure, and is only in the name of the employer. This is to reflect the fact that often this form of contract will be used by residential occupiers who may find it difficult to obtain insurance in joint names. Accordingly, the works will be uninsured unless clause 5.4A is also used, which obliges the contractor to insure the works.

2.2.5.3 NEC3 (core clauses 80–87)

The NEC3 approach to documenting parties' insurance obligations is less prescriptive than many other forms of contract. The specifics of the necessary insurances are entered into the contract data by the parties. For example, the contractor provides the standard insurances contained in the insurance table within clause 84, unless the contract data states that the employer is responsible for any of them. Details of any extra insurance the contractor or the employer are obliged to provide, should also be added into the contract data.

The benefits of this approach are that it is flexible to the needs of each project, and requires the parties to address and agree the insurance requirements before operating the contract. However, as the regime is less standardised, it can mean that the representatives of each party who operate the contract on an everyday basis, are not as familiar with who has insurance for each risk.

Each type of insurance is to be maintained in the joint names of the employer and the contractor, from the starting date until the defect's certificate, or termination if earlier. In this respect the NEC3 has more in common with the FIDIC rather than the JCT contracts.

NEC3 again follows the FIDIC approach of linking the insurance to the risks adopted by each party. The employer's risks are explicitly stated in clause 80.1 and any further risks to be allocated to the employer must be done so specifically by entries in the contract data. Clause 81.1 then states that the contractor's risks are the risks not carried by the employer. 'Risks not carried by the employer' could arguably include the compensation events in clause 60.1, as the employer takes the risk of additional time or money if any of these events occurs. This is something that the employer should take into account.

If any amount is not recovered from an insurer, then whoever takes responsibility for that risk under the contract shall be responsible for that amount. This would include either where an insurance policy is not in place, or where the specific terms of the policy prevent recovery, i.e. late notification.

The insurance table provides for:

- loss of or damage to the works, plant and materials (equivalent of JCT Option A, B or C insurance, however this does not cover plant);
- loss of or damage to equipment (not covered under the JCT regime);
- public liability (JCT clause 6.4.1); and
- employer's liability (JCT clause 6.4.1).

There is no requirement for professional indemnity insurance even if the contractor is to undertake design, or insurance of existing buildings where there are internal works. These can be entered in the contract data but the contract should ideally be amended to make specific provision for such insurances.

The policies must include a waiver of the insurer's rights of subrogation against directors and employees of the insured companies, unless issues of fraud arise. Subrogation rights give an insurer the right to step into the shoes of the insured party, e.g. a contractor, and pursue a claim against a third party in the contractor's name. However, construction contracts frequently contain waivers of the insurer's subrogation rights, against certain named or otherwise identified third parties. The third parties are generally parties identified under the same policy of insurance, and therefore the waiver can be of little practical consequence, as an insurer is less likely to bring a subrogated claim where it is providing the insurance.

The provisions in the NEC3 Short Form are simplified. As in the long form, the contractor is to provide the insurances that are listed in the insurance table. The insurance table in the Short Form also varies slightly. The insurance of the works is only required to be provided by the contractor until the completion of the works, rather than the expiry of the defects period. The remainder of the insurances are required to be maintained until the end of the defects rectification period.

2.2.5.4 FIDIC

The Red Book insurance regime is more prescriptive than the NEC3 equivalent and, in certain ways, more so than the JCT regime (clause 18.1) when addressing the general obligations to insure and the consequences of any failure to insure. The default position is that the contractor is required to procure and maintain all insurances unless agreed otherwise, or where insurance is not required. Whenever the contractor has to take out insurance, the employer must approve the terms of the policy.

The contract provides for three types of insurance.

Works, plant and materials (Clause 18.2): This insurance must be maintained in the joint names of the contractor and the employer, by the contractor, until the issue of the performance certificate. This is the certificate issued at the expiry of the defects rectification period. The equivalent insurance under the JCT contracts has only to be maintained until the start of the defects rectification period.

Another key difference to the JCT position is that the FIDIC terms provide that the insurance must be for the full reinstatement cost, including the cost of demolition, of removal of debris, professional fees and profit. These items and any other items which generally fall outside of the reinstatement costs must be included by the contractor in the reinstatement costs when procuring the insurance. The FIDIC terms also do not provide a percentage uplift of the reinstatement cost as in the JCT terms. Failure to take such issues into account when procuring the insurance can be costly.

The all risks insurance should cover risks other than (i) loss or damage to part of the works caused by the use/occupation by the employer of another part of the works, (ii) those risks which are not designated uninsurable and (iii) those risks included in the employer's risks clause. However, the exclusion does not include (c), (g) and (h) in this clause (clause 17.3), which must be covered by the insurance, if certain commercial conditions are met.

The insurance may exclude cover for defective design, materials or workmanship, but it must cover (i) loss or damage to other parts of the works as a direct result of that defect; (ii) loss or damage to the works arising from reinstatement of any part of the works as a result of that defect; (iii) loss or damage to any part of the works taken over by the employer as long as it is not the contractor's fault; and (iv) plant, equipment and materials not in the country of the works.

It should be noted that even though the contractor may be required to carry out design (clause 4.1), there is no obligation to procure professional indemnity (PI) insurance. A particular condition will need to be included in the contract if a specific PI insurance obligation is required.

Injury to persons and damage to property (clause 18.3): As above, the insurance must be maintained in the joint names of the contractor and the employer, against loss, damage, death or bodily injury to any physical property or person (except for those covered by clause 18.4), arising out of the contractor's performance of the works. As with clause 18.2 insurance, this must be maintained until the performance certificate is issued. To apply this insurance, specific provision must be made for it in the Appendix to tender.

There are certain exclusions from liability related to this insurance, such as where the damage is caused by an employer's risk (clause 17.3), where the damage is an unavoidable result of carrying out the works, or where the employer exercises his right to have the works executed on, over, under, in or through any land and to occupy such land for the works.

Contractor's personnel (clause 18.4): This insurance is intended to indemnify both the employer and engineer as well as the contractor, but it does not specify that it has to be in the joint names. It is to cover losses arising from the injury, sickness, disease or death of any person employed by the contractor or the contractor's personnel, but may exclude claims arising from the negligence of the employer or employer's personnel.

It is to be maintained during the period that these personnel are assisting in the execution of the works. Compared to the definitive duration of the works and public liability insurances, this is decidedly vague.

The Yellow Book terms are identical to the Red Book.

The insurances required under the FIDIC Short Form contract are similarly compatible, although the provisions are less detailed. If any amounts are received under the insurance policies, the amounts are to be held jointly by the employer and the contractor to remedy the loss or damage caused. There is no specified end date for when the contractor is no longer required to maintain the insurances. However, under the risk and responsibility clause (clause 13), responsibility for the works shall pass to the employer on the issue of the taking over certificate, so the contractor's obligation to maintain the insurance can be said to cease at this point. It is also possible for the employer to specify further types of insurance in the Appendix. It appears that the contractor will not be entitled to any payment from the employer for the remedial work required (clause 13.2). Therefore, if any monies received under an insurance policy do not cover the cost of the remedial work, then the contractor will have to bear the outstanding cost.

2.2.5.5 ACA/PPC 2000

To interpret the insurance provisions, four different sections of the contract must be considered:

1) The Project Partnering Agreement (Contract Particulars in JCT terminology) or the Commencement Agreement set out which members of the partnering team (at least employer and contractor, but also potentially other members of the supply chain) are required to procure insurance, the basis and the value of such insurance;

2) The Partnering Terms (Conditions of Contract in JCT terminology) set out the basic obligations of the party taking out the insurance; and

3) Appendix 4 sets out further specific requirements of each insurance.

4) There are seven types of insurance specified in the contract, not all of which are mandatory:

   - insurance of the works, goods, materials, equipment and existing structures including their contents (clause 19.1);
   - insurance similar to JCT clause 6.5.1 insurance (clause 19.1);
   - third party liability insurance, covering injury/damage to any person and/or property (clause 19.3);
   - professional indemnity or product liability insurance (clause 19.4);
   - environmental risk insurance (clause 19.5);
   - latent defects insurance (clause 19.6); and
   - whole project insurance (clause 19.7).

Whole project insurance, or integrated project insurance, is a single policy taken out by the client which covers all members of the project team. Each member of the team waives its rights to claim against other members of the team, and instead look to the policy to recover any losses. This is intended to prevent one party asserting blame against another and the risk of competing claims under overlapping policies. If the target cost is missed, then there will be an excess on the policy which will be payable in pre-agreed shares by the members of the team. This is similar to a 'pain share' mechanism under a target cost contract.

2.2.6 Latent Defects/Decennial Insurance

Decennial or latent defect insurance covers physical damage to insured premises caused by an inherent defect in the design, materials or construction of the structure. Typically, such insurance policies also cover the imminent threat of collapse arising out of an inherent defect. Insurance cover commonly lasts 10 to 12 years and commences on issue of the certificate of practical completion.

This type of insurance is not as widely used in the UK as elsewhere. The insurers offering such cover are limited in number, with the result that policy premiums and excesses can be high. Further, before an insurer will provide such insurance, it will have to satisfy itself that the construction of the building is generally sound. Security against defects in the UK is more commonly dealt with by the provision of collateral warranties backed up by professional indemnity insurance.

However, the costs and other limitations of this insurance should be balanced against the potential advantages. For example, there is no need for the policy holder to prove liability. If a latent defect arises, the policy provides immediate funds for remedial works.

2.2.7 Risk management: time, cost and quality

The risk management provisions in each of the contract forms discussed here should generally be read in conjunction with the insurance clauses for each contract. In some contracts, such as those in the NEC and FIDIC suites, there is a clear relationship between the risk and insurance clauses.

2.2.7.1 JCT DB and SBC

The later, more innovative forms of JCT contract have started to emphasise risk management. Procedures for dealing with errors and discrepancies in documents are legislated for and mechanisms for sectional completion and partial possession allow the employer flexibility to take over areas of the project on an 'as needed' basis. Relief events entitling the contractor to additional time and/or loss and expense are also specified as relevant events and relevant matters respectively (clauses 2.26 DB/2.29 SBC and 4.21 DB/4.23 SBC). Force majeure is a relevant event in the JCT suite, although not specifically defined. Therefore in assessing whether an event is an event of force majeure under a JCT contract, the event will need to be considered very much on its own facts applying common law principles unless expressly amended otherwise.

The sectional completion option is an integral part of each contract and is applied by making the appropriate entries in the contract particulars. Before a practical or sectional completion certificate is issued the employer may, with the contractor's consent, take possession of any part or parts of the works. The sections in possession of the employer are deemed practically complete and the defects rectification period for those sections begins (clauses 2.34 and 2.35). The insurance and liquidated damages provisions are adjusted accordingly.

The contractor has indemnity obligations in respect of costs, losses, claims and/or liabilities arising from personal injury, death or third party property damage. The personal injury indemnity is absolute, except to the extent others have caused or contributed to it (clause 6.1). The third party property damage indemnity applies to loss and/or damaged caused in the execution of the project and to the extent it is due to the contractor's negligence, breach of contract or statutory duty. Where the project is being undertaken in an existing structure, the indemnity does not apply to loss or damage to any property insured by the employer where such loss and/or damage is caused by an insured risk (clause 6.2 – specified peril exception where insurance Option C applies).

Relevant events are the relief events that allow the contractor to claim extensions of time and, similar to NEC3's compensation events, these events are broadly divided between employer and third party delay/acts of prevention, variation/change (and other employer) instructions, adverse weather and force majeure type events. Six of these events are also classified as relevant matters for loss and expense claims.

2.2.7.2 JCT IC, ICD and MW

The Intermediate forms contain the same relief event and indemnity provisions as the DB and SBC forms. There are also equivalent partial possession and sectional completion mechanisms. The Minor Works forms, as might be expected given their intended use for small works of limited duration, do not include mechanisms for partial possession and sectional completion. There is a general right to extensions of time for events beyond the contractor's control (clause 2.7) but no general or specific right to loss and expense. The indemnity obligation for third party property damage and personal injury are similar but not as extensive under the Minor Works forms as under the other contracts in the JCT suite.

2.2.7.3 NEC3

The risks designated as employer's risks (clause 80.1) can be supplemented by additional risks entered into the contract data; therefore the form of contract has a flexible approach to risk allocation which can be tailored to the specific constraints of a project. All risks not specifically designated employer's risks are allocated to the contractor (clause 81.1), therefore any additional employer's risks which the contractor may propose for the contract data are likely to be the subject of some debate between the parties. Employer's risks are compensation events (clause 60.1(14)). Each party indemnifies the other against the effects of events which are at its own risk (clause 83.1). Each party's indemnity liability is reduced if an event at the other party's risk contributes to any claim, proceedings or costs due as a result of the other event (clause 83.2).

Unforeseen events and acts of prevention are managed through the compensation event mechanism referred to in some detail in section 5. Physical conditions within the site (excluding weather conditions, which are provided for separately) are compensation events (clause 60.1(12)): 'Only the difference between the physical conditions encountered and those for which it would have been reasonable to have allowed is taken into account in assessing a compensation event'. In assessing physical conditions, the contractor is assumed to have taken into account the site information and any other available information (clause 60.2), therefore the allocation of risk in this context can be said to be reasonably balanced.

Force majeure is not referred to explicitly in NEC3, however the prevention clause (clause 19) and the corresponding compensation event (clause 60.1(19)) are essentially mechanisms for dealing with force majeure. The prevention clause is widely drawn, and so of concern to some employers. The language used is arguably sufficiently general to capture subcontractor breaches or acts of prevention which would not generally be considered force majeure events.

The contracts in the suite allow for sectional completion as an option, however the relevant provision (clause X5) must be expressly adopted in the contract data for the relevant project. NEC3 takes a different approach to the JCT as regards partial possession. The contractor does not have exclusive possession of the site, but is granted access to certain areas on *access dates* described in the contract data. The contractor must allow access by third parties to the extent specified in the works information. The employer is treated as having 'taken over' the works before completion, rather than having partial possession. Taking over is not fully defined by the contracts in the suite, but is linked to 'use' by the employer of the area taken over. Nevertheless, consistent with the JCT suite, the project manager must assess the benefit of taking over and make a proportionate reduction in the delay damages prescribed by the contract (if any).

Proactive risk management is central to the NEC philosophy. Risk is managed not simply through the allocation of specific risks, but also through the use of early warning procedures and risk reduction meetings. The risk register is a live contract document in the NEC3 suite and is intended to be updated in conjunction with the accepted programme as risks are identified and addressed (clause 16).

2.2.7.4 FIDIC

Responsibility for personal injury, death and damage to third party property is shared (clause 17). In line with the JCT and NEC3 suites, there are mutual indemnity obligations for loss or damage attributable to any negligence, wilful act or breach of each party (clause 17.1).

Neither party is liable to the other for indirect or consequential loss or damage and, subject to some exceptions, the contractor may cap its overall liability at a specified amount (clause 17.6). Until taking over, the contractor has full responsibility for the project (paragraph 1, clause 17.2) and during that time loss or damage to the works is at the contractor's risk and unless due to employer's risks (paragraph 3, clause 17.2). The contract identifies events and/or circumstances which are designated employer's risks (clause 17.3) in similar terms to NEC3, however unlike NEC3, the risks so designated are an exhaustive list. The list consists largely of force majeure type events and some events within the employer's control. The contractor is entitled to relief if it is delayed or incurs cost as a result of any employer's risks (clause 17.4).

The risk of the occurrence of exceptional events or circumstances is managed in its own provision (clause 19). The parties share an 'all reasonable endeavours' obligation to minimise any delay to the project arising from force majeure events and a duty to notify the other when a party ceases to be affected by a force majeure event. If the contractor is hampered by force majeure, he/she may be entitled to an extension of time and additional payment. As might be expected, the clause goes on to legislate for optional termination of the contract and release of the parties from their respective obligations if the force majeure event subsists for a continuous period of 84 days or a total of more than 140 days (clauses 19.6 and 19.7). FIDIC is not as prescriptive as NEC3 or PPC2000 about physical risks such as ground conditions or environmental hazards, therefore where these risks are significant for a particular project, consideration should be given to addressing them in the particular conditions.

2.2.8 Managing defects: quality

2.2.8.1 JCT DB and SBC

The contract particulars specify the rectification period within which defects, shrinkages or other faults in the works are to be made good (typically 12 months, although the JCT default position is six months; residential projects may require a longer rectification period of 18 or 24 months). In the DB form, the employer delivers the final schedule of defects to the contractor not later than 14 days after the expiry of the period (clause 2.35.2). The employer may however instruct rectification at any time during the period (clause 2.35.1). Defects are to be made good within a reasonable time and at no cost. The notice of completion of making good is when all notified defects have been completed (clause 2.36). In the SBC, the CA has the employer's role (clauses 2.38 and 2.39). The employer's certification role is in practice frequently delegated to the employer's agent.

2.2.8.2 JCT IC, ICD and MW

In keeping with the shorter forms of these agreements, the defects rectification obligations are similarly abbreviated. The default rectification period is three months, although, depending upon the nature of the project, a longer rectification period may be desirable. Unlike the DB and SBC forms the Intermediate forms do not appear to limit the make good defects obligation to the rectification period (clause 2.30).

2.2.8.3 NEC3

The NEC suite gives a narrower meaning to defect than the JCT (clause 11.2(5)). The test is whether the works accord with the works information, or if a design element, whether it complies with applicable law or accords with the design for the project which has been accepted by the project manager.

As might be expected given the suite's ICE origins, the clauses dealing with defects also cover testing and inspection. The applicable tests and inspections are those legally required or otherwise specified in the works information (clause 40.1). Where the project includes design and/or manufacturing and/or installation of complex items of plant or equipment, it will therefore be essential to ensure that the performance requirements and testing and inspection regime are dealt with comprehensively in the works information. Tests and/or inspections must be repeated if a defect is established (clause 40.4) and the contractor is responsible for the employer's costs in repeating the test or inspection, as assessed by the project manager (clause 40.6.). Plant and/or materials tested offsite cannot be delivered to site until the contractor has been notified that they have passed the relevant tests/inspections (clause 41.1).

Each party is obliged to notify the other of the type and timing of any such tests/inspections and the supervisor is entitled to witness the test and/or inspect the site prior to the tests (clause 60.2).

The rectification period (defects date) must be specified in the contract data Part 1 as a period of weeks after completion of the whole of the work. The Defects Certificate is issued at the later of the end of that period or the end of the last defects correction period (clauses 43.2 and 43.3 – defects correction time runs from the date of the relevant notice for defects identified during the works and from the completion date for those notified just prior to completion). The Defects Certificate may either identify uncorrected defects or state that there are no (patent) defects. Note that the defects correction period is the maximum time specified in the contract data within which a defect notified prior to the defects date must be corrected (clause 43.2).

The contractor may be instructed to search for defects at any time prior to the defects date and has a duty to notify any defects discovered during the duration of the project (clause 42). The contract permits the project manager to agree with the contractor that certain defects need not be rectified and to adjust the works information accordingly (clause 44). Notified defects which are not corrected within the prescribed time may be corrected by others and the cost reimbursed by the contractor (clause 45.1). The regime under Option C is substantially the same as under Option A, except that the cost of repeat tests/inspections is assessed on a different basis (clause 40.7).

NEC Short Contract does not include a testing and inspection regime pre- or post-project delivery, reflecting the intended use of the contract form for smaller less complex projects. The obligations in the contract to search for and notify defects are less exhaustive, but are ongoing, until the Defects Certificate is issued (clause 43). The contract assumes that the employer will draw defects to the contractor's attention (clause 40.1) and there is no option for certain defects to be omitted by agreement.

2.2.8.5 FIDIC

Both the Red and Yellow Books deal with tests on completion (clauses 7.4 and 9) and defects (clause 11) separately. Defects are not specifically defined in the contracts, the relevant provisions refer to 'defect or damage' and financial responsibility for them apportioned according to the cause of such defect or damage (clause 11.2).

Depending upon the nature of the project, tests may be required prior to completion. The engineer may attend the tests and can issue instructions varying the location or details of the required tests, including additional testing (clause 7.4 paragraph 3). The contractor is compensated for any delay or additional cost arising from such instructions. Similar relief is given to the contractor if tests on completion are disrupted by the employer (clause 10.3). Clause 9 legislates for the effects of delayed tests (clause 9.2) as well as for re-testing where the project or any part of it fails a relevant test (clause 9.3). Note that the engineer may still, at the employer's request, issue a Taking-Over Certificate if the tests on completion are not passed, subject to a reduction in the contract price (clause 9.4).

The Yellow Book goes on to prescribe the sequence for the tests on completion (which may be varied by amendments incorporated as particular conditions), briefly: pre-commissioning, commissioning and trial operation (clause 9.1). Any relevant performance criteria will need to be set out in the employer's requirements and comply with the schedule of guarantees. Trial operation does not constitute taking over. The contractor submits a certified report of the tests once passed to the engineer. If the works fail to pass the test on completion, and any subsequent re-testing exercise, the options are threefold: to order further re-tests, to reject the works or any relevant section, or if the employer so requests to issue the Taking-Over Certificate. Taking over on this basis triggers a proportionate reduction in the prices.

The defects notification period should be specified in the Appendix to tender. The contractor's obligations during this period are at its own expense to complete within a reasonable time any work outstanding at the date of taking-over and to remedy any defects or damage notified to it before the end of the defects notification period (DNP) (provided that the defect or damage can be attributed to any of the causes listed in clause 11.2; the causes are slightly broader than those describing a defect in NEC3). The DNP may be extended by up to two years if a defect is such that the project or any part (or a major item of plant) cannot be used until it is remedied (clause 11.3). Sanctions against the contractor for failing to remedy are similar to the JCT and NEC equivalents. The engineer issues a Performance Certificate after the later of (i) the expiry of the last relevant DNP and (ii) the delivery of all project documentation including test results.

Tests after completion are a feature of the Yellow Book and may be necessary for certain projects, particularly those involving complex plant and machinery. Physical outputs and performance criteria will need to be carefully defined in the contract to make the most of this process, including the schedule of guarantees which the contractor is prepared to offer. The tests are carried out as soon as practicable after taking over. The employer is required to give the contractor 21 days' notice of the tests, which must be carried out with 14 days of notification (clause 12.1).

If the contractor does not attend the tests s/he is deemed to have done so and accepted the results. The post-completion testing regime includes equivalent provisions for delayed and repeat testing (clause 12.3). Should the works fail to pass any post-completion tests, such tests are deemed passed if the contractor pays to the employer non-performance damages during the DNP. The contractor may instead propose to make adjustments or modifications to the works at a time convenient to the employer (where the contractor is delayed by the employer in gaining entry to investigate and make good, the employer is required to reimburse its costs plus reasonable profit). However, if the employer does not give the contractor reasonable access to carry out such work during the DNP, the works are deemed to have satisfied any post-completion tests and the contractor has no further rectification obligations to the employer.

FINAL THOUGHTS

Throughout this book, I have attempted to demonstrate the thought process that invariably every conscientious drafter of construction and engineering contracts will go through when reducing a commercial deal into a contractual form which provides certainty as to time, quality and cost to a project. It is not always easy, some risks can only ever be mitigated to one degree or another and more often than not (unless you are very lucky indeed) the odds are you will have to manage some kind of change to that originally planned or deal with something that has gone wrong.

We are lucky that the industries have over the years sought to come up with robust mechanisms for responding to such issues when they arise. Hopefully, it is now clear (even if it was not before) that there are a number of variables in play which means to say that you can only ever narrow down a draft by means of a number of critical items. The rest is important, sure, but not prescriptive.

The best contracts seek to respond proportionately to the fear of something going wrong (or someone making an honest mistake). There is often more than one way to respond to such matters as has been shown. My own experience over the past few years has been to see renewable energy schemes successfully procured using FIDC, IChemE (albeit, processes could have been streamlined for a particular AD plant procurement I was involved with) and even bespoke, funder-led EPC and EPCM contracts based on JCT wrap around for a 6.1MW/h rooftop solar PV installation. Provided the risk is dealt with in a sensible manner and the mechanics work, you can have the confidence that the standard form will respond in the way it was designed to do.

With any luck, this reference guide has provided a steer for those really in need and whom are tasked with sometime completing the impossible in a very short space of time. To those of you out there that perform these Herculean tasks, this book is meant for and dedicated to you.

Please feel free to contact me with your own experiences of drafting and operating a range of construction and engineering contracts at jon.close@icloud.com.

MORE BOOKS BY LAW BRIEF PUBLISHING

A selection of our other titles available now:

'Ellis and Kevan on Credit Hire, 5th Edition' by Aidan Ellis & Tim Kevan
'RTA Allegations of Fraud in a Post-Jackson Era: The Handbook, 2nd Edition' by Andrew Mckie
'A Practical Guide to Holiday Sickness Claims' by Andrew Mckie & Ian Skeate
'RTA Personal Injury Claims: A Practical Guide Post-Jackson' by Andrew Mckie
'On Experts: CPR35 for Lawyers and Experts' by David Boyle
'An Introduction to Personal Injury Law' by David Boyle
'A Practical Guide to Running Housing Disrepair and Cavity Wall Claims' by Andrew Mckie, Ian Skeate, Simon Redfearn
'A Practical Guide to Claims Arising From Accidents Abroad and Travel Claims' by Andrew Mckie & Ian Skeate
'A Practical Guide to Cosmetic Surgery Claims' by Dr Victoria Handley
'A Practical Guide to Chronic Pain Claims' by Pankaj Madan
'A Practical Guide to Claims Arising from Fatal Accidents' by James Patience
'A Practical Approach to Clinical Negligence Post-Jackson' by Geoffrey Simpson-Scott
'A Practical Guide to Personal Injury Trusts' by Alan Robinson
'Occupiers, Highways and Defective Premises Claims: A Practical Guide Post-Jackson' by Andrew Mckie
'Employers' Liability Claims: A Practical Guide Post-Jackson' by Andrew Mckie
'A Practical Guide to Subtle Brain Injury Claims' by Pankaj Madan

www.ingramcontent.com/pod-product-compliance
Ingram Content Group UK Ltd.
Pitfield, Milton Keynes, MK11 3LW, UK
UKHW020143290926
15471UKWH00004B/679

9 781911 035329